CHINESE URBANISM

This book provides a definitive overview of contemporary developments in our understanding of urban life in China. Multidisciplinary perspectives outline the most significant critical, theoretical, methodological and empirical developments in our appreciation of Chinese cities in the context of an increasingly globalized world. Each chapter includes reviews and appraisals of past and current theoretical development and embarks on innovative theoretical directions relating to Marxist, feminist, post-structural, post-colonial and 'more-than-representational' thinking. The book provides an in-depth insight into urban change and considers in what ways theoretical engagement with Chinese cities contributes to our understanding of 'global urbanism'. Chapters explore how new critical perspectives on economic, political, social, spatial, emotional, embodied and affective practices add value to our understanding of urban life in, and beyond, China.

Chinese Urbanism offers valuable insights which will be of interest to students and scholars alike working in geography, urban studies, Asian studies, economics, political studies and beyond.

Mark Jayne is Professor of Human Geography at Cardiff University, UK. He is a social and cultural geographer whose research interests include consumption, the urban order, city cultures and cultural economy and has published over 80 journal articles, book chapters and official reports. Mark is author of *Cities and Consumption* (Routledge, 2005), co-author of *Alcohol, Drinking, Drunkenness: (Dis)Orderly Spaces* (Ashgate, 2011) and *Childhood, Family, Alcohol* (Ashgate, 2015). Mark is also co-editor of *City of Quarters: Urban Villages in the Contemporary City* (Ashgate, 2004), *Small Cities: Urban Experience Beyond the Metropolis* (Routledge, 2006), *Urban Theory Beyond the West: A World of Cities* (Routledge, 2012) and *Urban Theory: New Critical Perspectives* (Routledge, 2017).

CHINESE URBANISM

Critical Perspectives

Edited by Mark Jayne

LONDON AND NEW YORK

First published 2018
by Routledge
2 Park Square, Milton Park, Abingdon, Oxon OX14 4RN

and by Routledge
711 Third Avenue, New York, NY 10017

Routledge is an imprint of the Taylor & Francis Group, an informa business

© 2018 selection and editorial matter, Mark Jayne; individual chapters, the contributors

The right of Mark Jayne to be identified as the author of the editorial material, and of the authors for their individual chapters, has been asserted in accordance with sections 77 and 78 of the Copyright, Designs and Patents Act 1988.

All rights reserved. No part of this book may be reprinted or reproduced or utilised in any form or by any electronic, mechanical, or other means, now known or hereafter invented, including photocopying and recording, or in any information storage or retrieval system, without permission in writing from the publishers.

Trademark notice: Product or corporate names may be trademarks or registered trademarks, and are used only for identification and explanation without intent to infringe.

British Library Cataloguing-in-Publication Data
A catalogue record for this book is available from the British Library

Library of Congress Cataloging-in-Publication Data
A catalog record for this book has been requested

ISBN: 978-1-138-20171-2 (hbk)
ISBN: 978-1-138-20172-9 (pbk)
ISBN: 978-1-315-50585-5 (ebk)

Typeset in Bembo
by Apex CoVantage, LLC

CONTENTS

ILLUSTRATIONS

Figures

Tables

CONTRIBUTORS

Xiaomei Cai is a Professor in the School of Tourism Management, South China Normal University, Guangzhou, China. Her research interests focus on hotel management and identity politics.

Jingfu Chen received his PhD from the National University of Singapore. His research interests focus on mobilities and politics. In particular, his recent work explores social interaction, equality and everyday practices of (im)mobilities in Chinese cities.

Gengzhi Huang is an Assistant Professor in the Guangzhou Institute of Geography, Guangdong Academy of Sciences, China. His research focuses on urban informality, migrants and labour geographies. He is directing the research programme *Development and Spatiality of Urban Informal Economies in China*, supported by the National Science Foundation of China (NSFC). His recent research on urban informality has been published in *Antipode*, *Geoforum*, *Urban Studies* and the top Chinese geographical journal *Acta Geographica Sinica*.

Alison Hulme lectures in International Development at the University of Northampton, UK. Alison specialises in the development of China as a consumer nation, alongside consumption, globalisation and material culture more generally. With a PhD in Cultural Studies from Goldsmiths, University of London, UK Alison has previously held positions at Royal Holloway, University of London, UK; University College Dublin, Ireland; and Beijing Foreign Studies University, China. In 2014, Alison was awarded the Ron Lister fellowship at the University of Otago, New Zealand. Alison has published widely in academic journals and edited volumes, and her monograph *On the Commodity Trail* was published in 2015. Alison is currently working on the economic history of contemporary China and completing her second monograph focused on frugality and thrift in historical contexts.

Mark Jayne is Professor of Human Geography at Cardiff University, UK. He is a social and cultural geographer whose research interests include consumption, the urban order, city cultures and cultural economy and has published over 80 journal articles, book chapters and official reports. Mark is author of *Cities and Consumption* (Routledge, 2005), co-author of *Alcohol, Drinking, Drunkenness: (Dis)Orderly Spaces* (Ashgate, 2011) and *Childhood, Family, Alcohol* (Ashgate, 2015). Mark is also co-editor of *City of Quarters: Urban Villages in the Contemporary City* (Ashgate, 2004), *Small Cities: Urban Experience Beyond the Metropolis* (Routledge, 2006); *Urban Theory Beyond the West: A World of Cities* (Routledge, 2012) and *Urban Theory: New Critical Perspectives* (Routledge, 2017)

Annamma Joy is Professor of Marketing at the University of British Columbia, Canada. Annamma's main research interests focus on aesthetics and consumption, including the consumption of art, luxury brands, fashion and wine. Annamma has published in journals such as *Journal of Consumer Research*, the *Journal of Consumer Psychology*, the *Journal of Retailing*, the *International Journal of Research in Marketing* and the *Journal of Consumer Culture*.

Ho Hon Leung received his doctoral degree in Sociology, McGill University, Canada. He is a Professor of the Department of Sociology, State University of New York at Oneonta (SUNY Oneonta, USA). His research interests include ethnic relations, immigration, urban studies, and comparative aging and his most recent work focuses on cross-cultural comparative study of cultural heritage and values in 'time-honored restaurants'. Leung is the Chair of the Center for Social Science Research and Academic Excellence Committee at his college. Ho Hon has published in national and international journals and has co-edited a number of books including *Imagining Globalization: Language, Identities, and Boundaries* (Palgrave Macmillan, 2009).

Yan Li is a PhD candidate in the Department of Public Policy, City University of Hong Kong. Her research focuses on urban governance in China with a specific focus on precarious creative labour in Xiamen. Yan holds a Master of Science degree from the Department of Anthropology at the London School of Economics and Political Science, UK.

Tao Lin is a researcher working on urban planning in the China Academy of Urban Planning and Design. Tao holds a PhD degree in Human Geography from Hong Kong Baptist University. His research focuses on the linkages between informal and formal economies and the governance of (in)formality. Tao's recent research has been published in the *Journal of Transport Geography*.

Chen Liu is a Lecturer in the School of Geography and Planning at Sun Yat-Sen University, China. Chen was awarded her PhD in cultural geography from Royal Holloway, University of London, UK. Chen's research is mainly focused on food consumption and popular culture in both domestic and public spaces in urban China.

Junxi Qian is an Assistant Professor in the Department of Geography, The University of Hong Kong. His research is located at the intersection of geography, urban studies and cultural studies. Junxi holds a PhD in Human Geography from University of Edinburgh, Scotland, UK and his recent research focuses on urban public space, place identities, the restructuring of Chinese cities, and China's ethnic frontiers. Junxi is the author of *Re-visioning the Public in Post-reform Urban China: Poetics and Politics in Guangzhou* (Springer, 2018) and has authored numerous journal articles and book chapters in both Chinese and English languages.

Melissa Y. Rock holds a dual-doctoral degree from The Pennsylvania State University's (USA) Departments of Geography and Women's, Gender, and Sexuality Studies. Melissa is currently an Assistant Professor of Geography, and an affiliate faculty member in the Department of Women's, Gender, and Sexuality Studies and Asian Studies Program, at the State University of New York (SUNY) at New Paltz (USA). In addition, Melissa serves as a co-founder and organizer of SUNY New Paltz's Digital Arts, Sciences and Humanities (DASH) Lab, which provides trainings and workshops on the integration of digital pedagogies and tools into teaching curriculum and research. Melissa is an urban and feminist geographer whose research centres on the intersection of geographies of care, social reproduction and urbanization in China.

Hyun Bang Shin is Associate Professor of Geography and Urban Studies in the Department of Geography and Environment at the London School of Economics and Political Science (UK). His main research interests lie in critical analysis of the political-economic dynamics of urban (re)development. Hyun has written widely on speculative urbanisation, the politics of redevelopment and displacement and urban spectacle. Hyun has co-edited *Global Gentrifications: Uneven Development and Displacement* (2015, Policy Press), has co-authored *Planetary Gentrification* (2016, Polity Press) and is currently guest-editing three journal special issues, one for *Urban Studies* on gentrification in East Asia and two for *Urban Geography* on gentrification in Latin America and mega-events in the Global South. Hyun is also working on a monograph entitled *Making China Urban* (Routledge), and a co-edited volume *Contesting Urban Space in East Asia* (Palgrave Macmillan). Hyun is a senior editor of the journal *CITY*, and also sits on the international advisory boards of the journals *Antipode*; *City, Culture and Society* and *China City Planning Review*.

John F. Sherry, Jr. is the Herrick Professor of Marketing at the Mendoza College of Business, at the University of Notre Dame (USA). John is the past President both the Consumer Culture Theory Consortium and the Association for Consumer Research. He has published extensively in top journals such as the *Journal of Consumer Research*, the *Journal of Marketing*, the *Journal of Retailing* and the *Journal of Consumer Psychology*.

Xiaobo Su is an Associate Professor in the Department of Geography, University of Oregon (USA). Xiaobo studies the dynamic relation between home and mobilities in contemporary China.

Jeff Jianfeng Wang is Assistant Professor of Marketing at the City University of Hong Kong. Jeff obtained his PhD in Marketing with a minor in Sociology from the University of Arizona (USA). Jeff's work appears in leading marketing journals such as *Journal of Academy of Marketing Science*, the *Journal of Retailing*, the *Journal of Advertising Research*, the *Journal of Business Research*, the *Journal of Public Policy and Marketing*, the *Journal of Consumer Affairs*, and the *Journal of Macromarketing*.

Jun Wang is an Assistant Professor at the Department of Public Policy of City University of Hong Kong. Jun's research interests include cultural/creative cities, heritage construction and recently, territoriality, citizenship and the precarity of the creative workforce. Jun has co-edited the book *Making Cultural Cities in Asia: Mobility, Assemblage, and the Politics of Aspirational Urbanism* (Routledge, 2016) and has authored papers in various journals, such as *Geoforum*, *Urban Geography*, *Territory, Politics, Governance*, and *Cities*. specialising in

Junfang Xie graduated from the Department of Landscape at the University of Sheffield, UK with a PhD focused on urban planning, policy and everyday use of green space in Shanghai. Junfang currently holds a post-doc research position in the College of Architecture and Urban Planning, Tongji University, China.

Desheng Xue is Professor of Human Geography and the Dean of the School of Geography and Planning, Sun Yat-sen University. Desheng is the vice-chair of the Geographical Society of China and he currently directs the National Science Foundation of China's *Globalizing Cities of China Since the Reform: Process, Pattern, Space and Dynamics.* Desheng's research focuses on world cities and globalisation in China and he has recently published in journals such as *Cities*, *Antipode, Geoforum* and the top Chinese geographical journal *Acta Geographica Sinica.*

Amy Yueming Zhang is a Lecturer in Human Geography at the School of Geography, Earth and Environmental Sciences, University of Birmingham, UK. Yueming is an urban and economic geographer with research interests focused on the political economy of urban land, the built environment, place-making, and policy mobilities. Yueming has published on issues of land politics and finance in Chinese cities as well as work considering the politics and aesthetics of China's changing urban arts districts. Yueming's current work is focused on 'creative cities' 'beyond the West'.

Jie Zhang holds a PhD from the Beijing Foreign Studies University, China. Jie's research focuses on critical theories, comparative literature and cultural studies, especially with regard to the cultural politics of train travel. Jie is a Lecturer at Hainan Normal University, China and is currently working on critiques of 'Western' urban theory and noisy cities.

J. J. Zhang is an Assistant Professor in the Department of Geography at The University of Hong Kong. He gained his PhD from Durham University, UK. His doctoral research focused on the cultural geo-politics of rapprochement tourism between China and Taiwan. J.J.'s research interests lie in the intersection of material culture, tourism and geopolitics and he writes on issues pertaining to bordering practices and social memories as they unfold as 'lived' and 'everyday'.

Yimin Zhao received his PhD from the London School of Economics and Political Science (UK). His research interests focus on the role of the Chinese state in urban development. In particular, his recent work explores land rights and urban green belts in China.

ACKNOWLEDGEMENTS

The editor would like to thank Faye Leerink, Priscilla Corbett and Ruth Anderson at Routledge for their patience and support, and all of the contributors for ensuring that the compilation of this book was an enjoyable process. Mark would also like to thank all at the 4C5M studio for organising the conferences that have encouraged his visits to China over the past few years, and Professor Cecilia Wong (University of Manchester, UK) and Professor Hongyang Wang (Nanjing University, China) for enabling a visiting fellowship to Nanjing. Special thanks to the numerous other colleagues and students who have become valued friends and WeChat companions as well as being welcoming and generous hosts during visits to Chinese cities. Mark would also like to express his gratitude for the opportunities to present his research to colleagues and students and to engage in rich and critical discussions about Chinese urbanism at The School of Architecture and Urban Planning, Nanjing University; The School of Chinese Language and Culture, Tianjin University of Technology; The Faculty of Architecture and Planning, Chongqing University; The College of Urban and Environmental Sciences, Central China Normal University, Wuhan; and The School of Geography and the Centre for Social and Cultural Geography, South China Normal University, Guangzhou.

And, last but by no means least, the editor gives a big cwtch to all of those who continue to offer kindness, criticism, company, fun, support, friendship and academic inspiration: David Bell, Gary Bridge, Noel Castree, T. C. Chang, Allan Cochrane, Neil Coe, Bethan Evans, Matthew Gandy, Sarah Marie Hall, Martin Hess, Stephen Hincks, Phil Hubbard, Sarah L. Holloway, Loretta Lees, the Leyshons, Andrew Mould, the MMU boys, Chris Perkins, Dr. and Mrs Potts, Jennifer Robinson, Kevin Ward, Andrew Williams, the Wollongong boys and Gill Valentine – Iechyd da!! Special thanks to First Class global traveller Daisy.

The authors, editors and publishers would like to thank copyright holders for their permission to reproduce the following material:

Jayne, M. and Leung, H. H. (2015) 'Embodying Chinese urbanism: towards a research agenda', *Area*, 46 (3), 256–267.
Xie, J. (2015) 'Landscape design, housing and everyday use of green space in urban China, *Geography Compass,* 9 (1), 42–55.
Reproduced by permission of John Wiley and Sons, London and New York
Zhang, J. J. (2013) 'Borders on the move: cross-strait tourists' material moments on "the other side" in the midst of rapprochement between China and Taiwan', *Geoforum*, 48: 94–101.
Reproduced by permission of Elsevier Publishing, London and New York.

1

AN INTRODUCTION TO CRITICAL PERSPECTIVES ON CHINESE URBANISM

Mark Jayne

It is surprising that the unprecedented scale and speed of urbanisation in China over the past two decades has not placed theoretical and empirical study of Chinese cities at the heart of contemporary critical urban scholarship (see Jayne and Ward 2017). Growing academic interest in Chinese urbanism around the world has nonetheless led to writing about Chinese cities mirroring the diversity of arguments, approaches, topics and case-studies that characterises international research agendas. So how can we explain this curious situation? Given the remarkable transformation of Chinese cities and the now voluminous, and ever growing, amount of research and writing, why is it that scholarship focused on urban China is not making more of an impact as cutting-edge, innovative and adventurous urban thinking? To what extent, and how, are Chinese urban scholars, and scholars of urban China responding to this lack of international influence? *Chinese Urbanism: Critical Perspectives* is the first collection of essays to ask, reflect on, and respond to these questions and challenges.

Urban theory and Chinese urbanism

In order to frame the contributions made by authors throughout this book this introduction begins by discussing the relationship between urban theory and Chinese urbanism followed by a summary of key theoretical debates and empirical topics addressed within and across the chapters. Beforehand, however, I make a brief detour to address what might well have been the first thing to spring to the mind of anyone with an interest in critical urban perspectives whose gaze was drawn to the title or cover of this book. Did you ask yourself how is it that an urban social and cultural geographer living and working in the small city of Cardiff, in the small country of Wales has come to edit a book that claims to advance our understanding of Chinese cities?

A rather convoluted answer to this question begins with regard to another small city: Stoke-on-Trent (also known as The Potteries) located in the English midlands, a city that is (perhaps ironically) best known in many parts of the world (historically, but arguably less so today) for its production of ceramics – inspired and enabled of course in no short measure through exploitation and replication of centuries of knowledges, technologies, inventions, techniques, expertise and designs pioneered in China. It was in Stoke-on-Trent, during my PhD research that I became more and more frustrated by the urban studies literature that I was reading which was overwhelmingly dominated by theoretical arguments, case-studies and empirical material generated in big cities throughout Europe and North America. The urban theory that I was reading can perhaps best be characterised as generalisable depictions of urban change proliferated through research undertaken in 'big' or global cities and 'the great' metropolises – so very different from my own case-study city.

London, Paris, New York, Berlin, Los Angeles, Tokyo, Las Vegas were used as templates for considering epochal urban conditions relating to Fordism/post-Fordism, industrialisation/post-industrialisation, modernism/post-modernism, global cities/global regions and so on. These global metropolises, and the spaces and places within them were invariably depicted as archetypal or paradigmatic in order to support the theoretical arguments being championed (see for example, Veblen 1899; Simmel 1903; Christaller 1933; Adorno and Horkeheimer 1947; Benjamin 1999; Hall 1966; Zukin 1989; Knox 1987; Castells and Portes 1989; Harvey 1989a and b; Soja 1989; Davis 1990; Sassen 1994; Beaverstock *et al.* 1999; Storper 1997; Scott 1998, 2001 and so on). Moreover, even when studies focused on urbanism beyond 'big cities', looking at national capitals, or 'regional' and 'provincial' cities, researchers tended to judge them, at best, as of 'secondary' importance, or at worst by defining them as 'failing'. Despite progress being made in rebalancing the woeful neglect of small cities by urban theorists through research which considers the urban world as not made up of a handful of global metropolises but characterised by heterogeneity, there is still much work to do to challenge urban studies orthodoxies that define small cities as of 'lesser' importance (Bell and Jayne 2006, 2009; Jayne *et al.* 2011; Jayne 2013).

Embedded in these theoretical concerns and arguments applied to championing the importance of studying small cities, my first steps towards critical intervention with Chinese urbanism emerged through inspiration and a mix of wonder and frustration engendered by my first visit to China. Excitement, confusion, bemusement, amusement, and at times fear, filled my senses. Bombarded by thrilling sights, sounds, tastes, smells and atmospheres – I found Chinese cities to be awe-inspiring! I really wanted to know more! I admit to my embarrassment however, that at that time I had read relatively little academic writing about Chinese cities. Indeed, seeking to make sense of the heady mix of my first exciting experiences of China's urbanism coupled with attempts to address my limited knowledge of academic literature relating to China, in parallel with a lack of understanding of the orthodoxies of knowledge production in China turned out to be a challenging process. My confusion was keenly felt as I sat watching a keynote presentation at the conference that I had travelled to China to attend. The lecture (presented and translated by the speaker in both Mandarin and English) was based on research related to the redevelopment of a historic urban quarter in one of China's largest cities and was a fascinating story, charismatically presented (including some brief pointers to comparative case-studies throughout the world); focused on aesthetics and design, government and 'private sector' partnership working, the displacement of long-established communities and emergence of new restaurant/bar based night-time consumption cultures. There was however, an absence of theoretical content throughout. In wishing to know more about the urban thinking that underpinned the presentation, during the question and answer session I quizzed the speaker by asking about the theoretical inspiration for the research and paper. While there may very well have been nuances to the answer I received that were lost in translation (from Mandarin to English), the response seemed to be that 'the case-study spoke for itself'. My follow up question regarding the usefulness of theories generated in cities in Europe and North America (relating for example to writing on gentrification, class, purification of space, consumer culture and so) met with a more straightforward response that such ideas were not applicable to China.

Following my bemusement at this response I was pleased to find that such ontological and epistemological framing was generally not replicated throughout the remainder of the conference papers, nonetheless what Nigel Thrift (2000) describes as persistent myths about cities did spring to mind that day. First, Thrift argued against the idea that cities are becoming globally homogeneous, and that diversity of supposedly serial, homogeneous sites such as shopping malls, office blocks or in this case redevelopment of historic urban quarters is often assumed rather than critically explored. Thrift highlighted the importance of fully taking into account historical, political, economic, socio-spatial and cultural contexts in order to highlight how apparently similar urban spaces and places, institutions and everyday life vary across space and time. Second, Thrift pointed to long-standing myths of urban exceptionalism, which Pow later excellently elaborates with regard to Chinese urbanism. For example, Pow (2011: 47) characterises writing about urban China as often falling into a trap of assuming political, economic, social, cultural and spatial urban practices and processes 'peculiar to China [which] have rendered the Chinese urbanization trajectory *more different* than similar from Anglo-American cities ... [and accepted rather than critiqued that the] Chinese state in particular is seen to respond to and/or create conditions and institutions that render urban China's experience as unique and exceptional'. Such insights also point to the importance of Nigel Thrift's (2000) assertion that 'one size does not fit all' when it comes to thinking about urbanism, that cities (wherever they are in the world) are not homogenised entities. Not all cities are the same and no one story can tell of 'the city', no matter what its size, shape, location or national context. Such theoretical work insists on the imperative to fully account for heterogeneity and to challenge exceptionalist accounts of urban practices, identities and processes in order to critically re-imagine 'the city'.

My interest in global academic knowledge production and disparities of theory building further flourished soon after my first trip to China. I began to read literature outlining the details of a growing recognition by urban theorists of the need to critically engage with the theoretical, empirical and methodological challenges bound up with accounting for the diversity of cities around the world. Indeed, at that time there was a growing amount of writing being published which focused on the ways in which urban theory has tended to remain stubbornly focused on a small number of 'western' cities that act as the template against which all other cities are judged (see for example, Chakrabarty 2000; Connell 2007; Robinson 2006; Parnell and Robinson 2012; Isoke 2013; Roy and Ong 2011; Edensor and Jayne 2012). It was also being acknowledged by numerous scholars that this weakness in urban studies literature was further exacerbated by a lack of exchange of ideas and theoretical dialogue with 'non-western theory' (see for example, Edensor and Jayne 2011; Jayne 2013).

For example, Amin and Graham (1997) and later Robinson (2002, 2006) argued that all cities should be theorised as 'ordinary' – including those left 'off the map' of international urban research agendas. This argument makes clear that all cities are important centres (some arguably more than others) of globalising political, economic, social, cultural and spatial dynamics and that there is a need to look at a broader range of practices, processes, identities and autonomies that constitute structural and everyday urban experiences. Moreover, theorists including Chakrabarty (2000), Robinson (2002) and Connell (2007) highlighted the ways in which colonial relationships and imaginations underpin persistent asymmetrical relationships across international scholarship. While 'western' theorists more often than not ignore 'non-western' thinking scholars throughout the world are nonetheless expected to frame their work within 'an authorised western canonical literature' in order to publish in English language journals and to secure contracts with international book publishers. Robinson (2002) reminded urban theorists to acknowledge their own cultural and academic situatedness, and in doing so challenged us all to contribute to ensuring that 'western' academic publishing is more accessible to scholars around the world in order to counter a 'knowledge production complex'.

It is also important to note that work to de-centre urban theory from orthodoxies generated in Europe and North America and rallying calls to advance more heterogeneous, diverse and cosmopolitan urban studies have emerged and been acted on alongside proliferation of pluralistic theoretical approaches to understanding cities and urban life (see Jayne and Ward 2017). However, despite the growth of interest in Chinese urbanism there has been a relative lack of engagement with critical theoretical perspectives – including Marxist, feminist, post-structural, post-colonial, and 'more-than-representational' thinking – and the diverse theoretical debates and work that these voluminous bodies of literature have inspired and which has been vital in enlivening urban studies over the past few decades. Given the complexity of contemporary cities, it is not surprising that urban theory has become characterised by complex and heterogeneous thinking.

For example, as highlighted earlier in this chapter writing inspired by post-colonial theory has shown that researchers have only just scratched the surface of urban experiences around the world and there is need for a more diverse, inclusive and comparative urban theory (see for example, Bell and Jayne 2006, 2009; Robinson 2006; Parnell and Robinson 2012; Isoke 2013; Roy and Ong 2011; Edensor and Jayne 2012; Ward 2010; Jayne 2013). Directly emerging from this writing has been a return to comparative approaches (see Robinson 2006; Nijman 2007; Dear 2005; Legg and McFarlane 2008), such as Kevin Ward's (2009: 406) call for theorists to take seriously a 'relational comparative' approach which stresses interconnected urban trajectories to identify:

> how different cities are implicated in each other's past, present and future which moves us away from searching for similarities and differences between two mutually exclusive contexts and instead towards relational comparisons that uses different cities to pose questions of one another

Other recent innovative work seeks to develop a better understanding of territorial urban spatial fixes of capitalism while attending to increasing understanding of relational comparative urbanism (see McCann and Ward 2010, 2012; Jayne *et al.* 2011, 2013).

Alongside (and at times intersecting with) post-colonial thinking, and theoretical work emerging from its critique are of course other foundational debates that have been at the heart of urban studies. Of particular importance is Marxist and post-structuralist theory. Key figures in advancing what is a long tradition of Marxist urban thought include theorists such as Henri Lefebvre and more recently David Harvey. The illustrious career of Henri Lefebvre, a French sociologist and philosopher, began to flourish with his 'critique of everyday life' written in the 1930s. Working on similar theoretical and empirical terrain to members of the historic Frankfurt and Chicago School's of urban studies, Lefebvre considered topics such as alienation, boredom, consumption, and the promises of freedom that emerged with increasing leisure time across urban societies. Lefebvre suggested that work and leisure rhythms of capitalism were responsible for reducing self-expression and leading to a poorer quality of everyday life – in sum, that by colonising everyday life industrial capitalism was able to reproduce itself. Importantly, Lefebvre's (1984) arguments advanced previous theoretical debate through his focus on everyday life. Lefebvre highlighted that it was by championing progressive conditions of human life, rather than seeking to gain control of capitalist production, that humans could work towards achieving a socially just world. Lefebvre's (1968, 1991) influential later work from the 1960s onwards focused on the 'social production of space' and the 'right to the city', has also subsequently had a profound influence on urban theory. Of equal importance to the Marxist urban thinking of Lefebvre has of course been the work of David Harvey. Injecting critical perspectives initially into urban geography, and then proliferating influential thinking about cities across the social sciences David Harvey and those 'metroMarxists' inspired by his work have advanced our understanding of the profound influence of capitalism on urban experience (see Merrifield 2002).

In a similar vein, post-structural thinking inspired by theorists such as Foucault (1997a), Deleuze and Guattari (1987), Haraway (1991) and Latour (2005) has also had a profound influence on urban studies. For example, one thread of post-structural theory has inspired assemblage thinking which has been argued to have 'changed urban research' through a focus on topics such as materialities, (non)human actors, networks, policy, practices, ideas, learning, atmospheres (see Farías and Bender 2009; Blok and Farías 2016; Farías 2016). Furthermore, following a 'cultural turn' across the social sciences a growing number of urban theorists have considered emotions, embodiment and affect, or 'more-than representational' ideas in order to address mundane everyday practices that shape our social lives in particular sites (Thrift 2004b).

These (and other) innovative theoretical approaches have nonetheless left critical urban thinking stubbornly dominated by writers in Europe and North America. Despite implicit and explicit intentions to address this imbalance, 'it is the cities of the North which we have had in mind while writing the book' but there is hope that 'new perspectives' can be subsequently 'explored by others' (Amin and Thrift 2002: 3) – or that key arguments 'are relevant to the study of non-western cities too' (Hubbard 2006: 248). Indeed, despite the wealth of innovative, ground-breaking and rich theoretical resources now available, engagement with critical thinking nonetheless remains marginal in Chinese urbanism when compared to dominant economic and spatial analysis (although see the work of Kenworthy-Teather 2001; Gui et al. 2009; Kong 2011, Crang and Zhang 2012; He 2013; Wang *et al.* 2013; Pow 2012; Zhu and Wei 2016 – and of course the work of those who have contributed to this volume).

While it is clear that uptake of critical urban thinking has limited the ability of Chinese urban scholarship to make cutting-edge contributions to international research agendas we still need to ask how we can assess the growing amount of academic writing that is offering important insights and critical perspectives on the changing nature of cities and urban life in China. Empirically, attention has focused on urban and regional growth, land use and housing, the 'urban – rural' divide, physical and infrastructural restructuring, economic development, migration and social change (see for example; Logan 2001; Wu 2009; Lu 2011; Wu and Gaubatz 2012; Shaqiao 2014). A number of key texts have been at the vanguard of understanding Chinese urbanism with regard to a diverse range of urban practices and processes such as globalisation and economic reform, spatial and physical transformation (Xu 2000; Logan 2001; Wu 2009; Sit 2010; Wu and Gaubatz 2012; Ren 2013; Shaqiao 2014). Other volumes have considered topics such as urban design (Chen and Thwaites 2013); city and regional planning (Yu, L. 2014); housing inequality (Huang and Li 2014); modernity and space (Lu 2011); climate change and governance (Mai and Franchesh-Huidobro 2014); cinematic urbanism (Braester 2010); politics of community and social life (Heberer and Gobel 2013; Tang and Parish 2000); young people (Cockain 2011) and consumer culture (Yu, L.2014).

Important critical reflections on developments in Chinese urbanism are outlined by He and Qian (2017) in an insightful review of around 2500 articles published in international journals from 2003 – 2013 from across the disciplines that make up Chinese 'urban studies' – area studies, interdisciplinary social science, geography, sociology, planning development, demography, anthropology, political science, public administration, cultural studies and urban studies. Popular topics included the hukou system, sustainability and ecosystems, real estate development, divisions of labour, gated communities, migration, production networks, and wage inequality. Key themes discussed in these papers include globalisation, market transition and global cities, urban poverty and socio-spatial inequality, and rural migrants' urban experience. He and Qian (2016) also point to emerging frontiers of research relating to urban enclaves and spatial fragmentation, public space, consumption, middle-class urban culture, social movements, activism and the 'right to the city' as well as arguing the need for more scholarship on topics such as transnationalism, mobilities, gender, sexuality and so on. In applying a critical perspective to Chinese urban literature over the past decade or so He and Qian (2017) celebrate the breadth and diversity of scholarship, and rightly foreground areas of theoretical innovation and fruitful new avenues of research. However, in reflecting on progress to date He and Qian (2017: 463) nonetheless lament that Chinese urban scholarship can overwhelmingly be characterised by 'empirical studies and econometrical modelling' and go on to argue that, 'scholarship in urban China studies [has been] … dominated by a positive approach and generally lacked nuanced analysis and theoretical debates'.

While this critical interrogation of the strengths and weaknesses of Chinese urban literature is timely and pertinent, it also is worth noting that the weight of empirically focused and/or positivist ontologies and epistemologies is not however unique to Chinese scholarship and that around the world

> for a significant proportion of academics who write about cities, advancing or contributing to the development of urban theory is a minor or at best secondary concern. For example, the emphasis of many top ranked urban studies journals that publish research and writing on cities is often weighed towards presenting empirical descriptions and explanations of practices and processes in cities … Many academic writers engaged in working beyond universities, involved in policy and practice may also not consider themselves to 'do' theory at all.
>
> *(Jayne and Ward 2017: 3–4)*

Indeed, prior to the 1960s, and prompting more recent accelerations in the development and proliferation of pluralistic theoretical impulses, the relevance of urban theory in identifying and seeking to address inequalities and injustice in cities in Europe and North America had received significant criticism (see Harvey 1989a; Thrift 1993; Castells 1997). The hard work to overcoming these historic shortcomings has of course led to the kind of rich and vibrant thinking that now characterises contemporary urban theory. Indeed, for many academics it is the belief that urban theory is the most profound contribution to knowing, finding out about, and working to help change the world in progressive ways that makes academic engagement with cities unique. As Jayne and Ward (2017: 16) highlight, at the heart of innovative and cutting-edge academic work is urban theory that should be considered as

> the lifeblood of the academic system, the tools of the trade through which complex ideas and understanding are generated, expressed and represented. Thought of in this way, theory in academia is equivalent to the language, lexicons and vocabulary used in many work places around the world … It is theory that enables the expression and performance of critical thought and intellectual advancement, important not only for the production of academic knowledge, but also in influencing how academics find out about cities through the methodologies they use, and furthermore in work which seeks to positively influence diverse audiences beyond universities

When considering the role of urban theory within Chinese scholarship it is important not to forget the significant critical writing by Chinese scholars, not published in English language journals and books, or not directly engaging with urbanism *per se* which offer rich and detailed understanding of social and cultural life in China including a diverse range of topics, case-studies and contexts. However, while it is important to critically reflect on the strengths of Chinese urbanism it is also important to seek to understand and overcome weaknesses of academic knowledge production in China that lead to the dominance of approaches that are 'empirically focused … generally lacking nuanced … theoretical debates' (He and Qian 2017: 463). Indeed, this is a pertinent time to remind us all that academics do not work in a 'vacuum' or the over-used image, an intellectual 'ivory tower'. As Jayne and Ward (2017: 16) highlight 'urban theory, as it has always done, takes as its focus, its stimulus, the world around us'. If critical Chinese urban thinking and theorists are to take centre stage in leading contemporary and future international debates then political, intellectual and financial support within and beyond universities in China and elsewhere in the world is a necessity.

With this challenge at the forefront of our minds, and as this section of the chapter draws to a close it is perhaps timely to reflect on what kind of theory is best able to advance Chinese urban scholarship, and scholarship of cities in China for the twenty-first century. Theorists such as Chakrabarty (2000), Connell (2007), Robinson (2006), Parnell and Robinson (2012), Isoke (2013), Roy and Ong (2011), and Edensor and Jayne (2012) have rightly pointed to the failings of urban theory generated in Europe and North America and the problematic claims to universality despite specific geographical, historical and cultural origins. In parallel with such insights is the problematic acknowledgement that 'western-centric' theory either marginalises or subsumes 'non-western ideas', codifying and contextualising those ideas from a 'western' perspective, or more typically, ignoring them entirely, as if all cities around the world are awaiting interpretation for the first time. By foregrounding ideas that have emerged from study of cities around the world theorists such as Chakrabarty (2000), Connell (2007), Robinson (2006); Parnell and Robinson (2012), Isoke (2013), Roy and Ong (2011) and Edensor and Jayne (2012) thus call for all theoretical interventions, no matter where they emerge, to be taken seriously as a resource in aiding understanding of cities and urban life. Taking a different approach other theorists are importantly undertaking ways of 'doing theory' that do not implicitly draw on European and North American traditions of knowledge production, genealogies of ideas, key thinkers and writing conventions and so on (see for example, Simone 2004, 2001). Ward's (2009) view that comparative urban research must be attuned to 'theorizing back' in order to offer reflection on the geographically uneven foundations of contemporary urban scholarship also offers fruitful avenues for future research, as does Connell's (2007) challenge for scholars to become adventurous, and open to 'dirty theory' as part of an omnivorous approach that applies ideas that suit particular situations and contexts. All these approaches should offer inspiration to those of you who are determined to advance our understanding of cities in China and to locate Chinese urban theory at the heart of international research agendas.

He and Qian (2017: 473) champion such theoretical innovation and argue that 'Chinese urban scholars need to draw from multiple strands of philosophical thinning, both indigenous and western, instead of taking European-American theories as given … [and go onto ask] What can Chinese urban studies offer to destabilize conventional concepts and theories in urban studies and cognate disciplines … [and to] contribute to the development of urban theories that emerge from the Global South?'. Crucial to responding to this challenge is the need to acknowledge polyvalent and polydiscursive characteristics of theory from around the world that invariably emerge from fluid, multiple and dynamic arguments and debates rather than as part of a dominant genealogies of urban thinking. While this agenda is by no means easy as Jayne and Ward (2017: 17) remind us:

> no field of urban theory is pre-configured, waiting to be discovered … boundaries are made and remade through books like these for example or through plenaries at academic conferences or lecture courses in universities around the world. And, over the years, its edges have been reworked, subject to contestation and refinement from inside and from outside. What has emerged is an inter-disciplinary, internally-heterogeneous, multi-methods-based set of contributions under the auspices of urban theory

The chapters of this book thus build on the successes made to date and in doing so open up new rich and fruitful avenues towards the aim of ensuring that Chinese urbanism is at the heart of critical urban thinking for the twenty-first century.

Towards that goal, this edited collection includes the most important and innovative current critical contributions to understanding cities and urban life in China. The chapters are of course, a snapshot of the progressive direction of travel, an important milestone in advancing thinking, research and writing on Chinese urbanism. The authors outline critical interventions and discuss tensions within Chinese urbanism that are generative and productive, offering intellectual impulse for the future reworking of urban theory and its boundaries with regard to understanding

Chinese cities and urbanism around the world. Following the wonder and frustration engendered by my first visit to China outlined at the beginning of this section, I have attended conferences and been invited to take up visiting research fellowships, and present guest-lectures in cities including Nanjing, Macau, Hong Kong, Shenzhen, Zhuhai, Cixi, Suzhou, Tianjin, Chengdu, Chongqing, Wuhan, Shanghai, Sanya and Guangzhou. During these visits, I have enjoyed engaging in critical debate with inspiring colleagues along with wonderful hospitality including delicious food, wine and beer. While empirically my own research in China to date has been limited to a study of Wuhan/Manchester's city twinning activities (see Jayne *et al.* 2011, 2013); dancing in public space and the role of massage in Chinese cities (Jayne 2013; and see Chapter 14, this volume) I am very hopeful that I will be able to work with my Chinese colleagues and scholars of urban China on numerous research projects in the coming years. An urban social and cultural geographer living and working in the small city of Cardiff, in the small country of Wales has a lot to learn from the challenges and opportunities of undertaking research in urban China, that I am sure will significantly enrich my own critical urban imagination. I hope, at that the very least, that this edited collection and the critical perspectives throughout will convince you (if you are not already convinced) that Chinese cities are awe-inspiring.

Chinese urbanism: critical perspectives

This book provide an original and agenda-setting contribution to Chinese urban theory. Chapters outline the most significant critical theoretical, methodological and empirical developments in Chinese urbanism in the context of an increasing globally inter-connected world at the beginning of the twenty-first century. The authors offer theoretically informed argument and debate that raises important issues about heterogeneous cities throughout China. Each chapter includes critical and accessible reviews and appraisals of past and current theoretical development and/or embarks on innovative theoretical directions relating to Marxist, feminist, post-structural, post-colonial, and 'more-than-representational' thinking. The *critical perspectives* are organized into five sections; space and place; identity, lifestyle and forms of sociability; consumption and urban cultures; (im)mobilities and materialities; and bodies, emotions and atmospheres that when read separately and together present cutting edge contemporary urban thinking.

Part One focuses on space and place as key to advancing critical perspectives on Chinese urbanism. Focusing on locations and contexts such as public squares, neighbourhoods, parks, tea-houses, bars, nightclubs, universities, playgrounds, shopping malls, beaches, living rooms and bedrooms, suburban golf courses and so on – not as 'bounded' but connected and relational, we highlight that space and place are not passive backdrops but active constituents in Chinese urbanism. Chapter 2 by Junxi Qian addresses the fascinating topic of public space in Chinese cities. Junxi argues that the notion of public space, rooted in Greco-Roman legacies and later furnished in the European intellectual enterprise of describing the epochal transition to capitalist modernity, is underpinned by particular political and civic ideas that are not necessarily present in Chinese cultural and intellectual traditions. In order to theorise conceptions of public space and understand everyday practice in public space Junxi outlines two critical perspectives: first, that to view urban public space in China as separate from, and even antithetical to the state is not tenable in a Chinese context; and second, that despite state intervention to proactively shape public life and configure the state – society interface, China is far from bereft of spaces of genuine grassroots agency. Junxi's argument is developed with empirical evidence of embodied and performative practices of urbanites in post-reform China as they negotiate rapid social changes, through unfolding pleasures, social relations and subjectivities.

In Chapter 3, Hyun Bang Shin and Yimin Zhao critique sets of urban knowledge and practice that produces urban inequalities and injustice in contemporary China with reference to green belt policy in Beijing. Drawing on Lefebvre's (2003) consideration of 'official urbanism', the chapter reflects on institutional and ideological urbanism as a state project, integrating economic and political practices, which play a critical role in sustaining state strategies of land-based accumulation. Shin and Zhao also illustrate official urbanism instilled into national identities as a new and desirable way of life, an expression of the politics of urban space and place, with the Party-state's ideological, economic and political ambitions centre-stage. Following a focus on 'official urbanism', Chapter 4 looks at relational connections between 'the home' and public 'green space'. Junfang Xie discusses historic and contemporary relationships between landscape design, housing and green space, by drawing together writing focused on theorising nature and everyday life. The chapter offers insights into urban political, economic, social, and cultural change in Chinese cities and considers how critical perspectives at the intersection of thinking on 'nature' and 'everyday life' can inform socially progressive urban development, planning, policy and politics.

Part Two focuses on identity, lifestyle and forms of sociability as key elements in the ordering and experience of everyday life in Chinese cities. Chapter 5 considers female sex workers' everyday life in Dongguan, China's so-called sin city. Xiaomei Cai and Xiaobo Su advance long-standing interests of urban theorists with 'strangers' and 'encounter'. Thinking through 'stranger-ness' as an essential element of urban societies Xiaomei and Xiaobo highlight socio-spatial inequalities that people on the margin encounter in their daily life, and consider the complex and intersecting ways in which power operates through these encounters in Chinese cities. Chapter 6 by Alison Hulme looks at how Chinese urbanism intersects with ChinaNet. Focusing on environmental protests in Chinese cities Alison points to a balance between individuality and principles, encouraging and facilitating environmental concerns, and by unpacking specific campaigns and protests, the chapter argues against depictions of young people in China as being overly concerned with individual consumption and as being straightforwardly self-interested. Chapter 7 by Melissa Rock considers caring and community in everyday practices in Beijing. Melissa develops critiques of neo-liberalism, as well as feminist theory, to argue that while rapid urbanisation continues to facilitate increasing geographic proximity and population density, neoliberal commodification and privatisation combined with discourses of personal responsibility, modernity and embodied quality, or *suzhi* have succeeded in eroding the overlapping public – private interstitial spaces that have fostered intimate neighbourhood encounters and interactions. This chapter elucidates the ways in which Beijing's economic and spatial reforms frame and constrain socio-spatial practices of caregiving and community engagement.

Part Three includes chapters that challenge theorisation of consumption and urban cultures overwhelmingly dominated by studies in Europe and North American (see Jayne 2005). Chapter 8 by Annamma Joy, John Sherry and Jiangfeng Wang explores practices of tasting, savouring, and signalling through which luxury brands are embraced as emblems of status and refinement. The chapter draws on theoretical discussion of 'distinction and taste' to explore temporal, spatial and social components of urban Chinese consumers' desire for style and modernity exemplified by Western brands, and for accumulating professional and social capital through luxury consumption, as practised by the affluent, nouveau riche, and aspirational Chinese upper-middle class. In Chapter 9, Chen Liu takes us from public and commercial spaces such as street markets and shopping malls to consider consumption at 'home'. Focusing on food (and drink) spaces, settling, storage and wasting Chen develops theoretical engagement with 'accommodating' and 'dwelling' to highlight the importance of understanding domestic consumption as a reflexive and critical narrative that signifies agency and individualisation as an experience of emancipation from the de-commodified past in urban China. In Chapter 10, Amy Yueming Zhang takes us back to the topic of urban regeneration in Chinese cities. Focusing on Beijing Design Week, Yueming offers important critical perspectives on the role of art and culture in urban regeneration by exploring the class dimension of urban politics in China.

J. J Zhang's contribution begins Part Four, with all the chapters in this section considering urban (im)mobilities and materialities as key to understanding the contemporary urban experience. In Chapter 11, J. J. interrogates the 'urban cross-border mobilities' of Taiwanese tourists in China by examining travel experiences at border areas through the lens of 'materialities' and 'liminality'. J. J. argues that the edges of cities are important in understanding Chinese urbanism and that border areas are not empty spaces of transition, but filled with identity negotiations and performances. This chapter advances theoretical consideration of 'mobilities', 'materialities', 'liminalities' and 'play', to offer timely reflections on urban China in an increasingly interconnected world. In Chapter 12, Jingfu Chen unpacks rhythms and (im)mobilities and the 'slow life' in the small city of Sanya. In the context of the dramatic socio-economic reconfiguration and modernisation in Sanya, Jingfu shows that 'rhythmanalysis' is a vital theoretical resource to consider how traditional ways of living appear out of step and are considered as problematic by many people, despite being enjoyed by many residents. In Chapter 13 Gengzhi Huang, Tao Lin and Desheng Xue focus on urban informality and migrant workers' negotiation of work/life balance. Based on a critical reading of current perspectives on informal economies, this chapter explores migrant workers' motivations for participating in informal work and the concomitant meaning for them in China. This chapter calls for more research on the articulation of migrants' informal working practices and their everyday life to better understanding the meaning of informality in Chinese urbanisation.

Part Five, the final section of the book, includes chapters which consider 'more-than-representational' theories of embodiment, emotions and atmospheres which have offered much to enliven urban studies over the past few decades. In Chapter 14, Ho Hon Leung and myself focus on embodied urban geographies. More specifically we consider case-studies of public dance and massage in order to show how critical engagement with everyday social and cultural forms and practices has much to offer understanding of urban life in China. Discussing theoretical terrain relating to public/private space, individual/collective practices and experiences and comfort/discomfort, we signpost important contributions that critical research into Chinese urbanism can make to broader debates within and beyond urban studies. Chapter 15 by Jie Zhang considers noisy cities by focusing on urban soundscapes and the spatial-temporal nature of 'noise' in cities as an increasingly important issue of academic, policy and popular concern. Jie advances critical perspectives on soundscapes, which is a bourgeoning and exciting area of urban studies, by focusing on neighbours, household appliances and traffic to highlight the paradox of 'noisy' city life with regard to embodied, emotional and affective urbanism. In Chapter 16, Jun Wang and Yan Li take us back to the topic of creativity but this time to look at 'moral atmospheres'. By deploying the concept of affective atmosphere, the chapter discusses the city of Lishui, where traditional adoption of the French Barbizon School of painting has been a central part of tourist-led urban regeneration schemes. Jun and Yan argue that such strategies cannot be understood simply as a top-down state-sanctioned image imposed on the society, but also is felt, practised and embodied by subject-citizens indulging in this atmosphere. Jun and Yan thus question how moral images are challenged in the face of concerns of political passivity, and hegemonic and anti-hegemonic actions.

Chapter 17 highlights key themes within and across the sections and chapters of the book. This brief Afterword offers insights into the opportunities and challenges of advancing critical Chinese urbanism for the twenty-first century.

PART I

Space and place

2

TOWARDS CRITICAL URBANISM

Urban public space in modern China

Junxi Qian

Introduction

In an essay on urban public spaces 'beyond the west', elsewhere I have argued that the notion of public space, rooted in Greco-Roman legacies and later furnished in the European intellectual enterprise of describing the epochal transition to capitalist modernity is underpinned by specific political and civic *ideals* (Qian 2014a). On the one hand, public space stands for a democratic forum that celebrates the unfettered communication of ideas and claims. On the other hand, public space may be conceived of as a civic arena whereby identities and differences are rendered visible, and thereby acknowledged and equalised. Nevertheless, in Qian (2014a) I also contend that beyond its Anglo-European underpinning, the centrality of public space to social formations is by no means eclipsed by the absence of normative ideals in cities 'beyond the West'. Thus, we need to attend to the ways in which context-contingent conceptions of *publicness* are assembled via situated discourses, practices and social relations, at the same time as understanding how public spaces 'beyond the West' are 'extensively used and appropriated, as well as profoundly politicised and contested' (Qian 2014a: 874). In the Chinese context, for example, the use of public space for making political claims and promoting the acknowledgement of cultural difference has been widely observed. However, I argue that such insights are not always explicable in terms of normative political and civic ideals and indeed that everyday life and practices ensure that public spaces are more dynamic and fluid than is often depicted. Such insights highlight the importance of sharpened sensitivities to local contexts and geographical and historical specificities.

This chapter builds on these theoretical premises in order to critically reflect on the making of public space in modern Chinese cities.[1] By 'modern' Chinese cities, I refer to a period that spans from the late Qing Dynasty (the late nineteenth to early twentieth centuries) – a time when, caught in turbulence of European-American imperial expansion China witnessed inchoate industrial and urban modernity – to the post-Economic-Reform present. While the association of public space with liberal democracy is far from definite in Chinese cultural and intellectual traditions, I highlight the importance of addressing the extent to which Chinese society 'idealises' public spaces in geographically and historically contingent ways. For example, writers have highlighted how public space in China has been elevated to the status of material incarnation of new thoughts, values and ideologies; glorified, variably, as an iconographic landscape for expression and resistance (Lee 2009); depicted as the harbinger of Western (progressive) urban civilisation (Lee 1999); and has been considered as a beacon of social progress and reform (Shi 1998). However, this chapter underscores that urban publics are not merely abstract ideological constructions, but enlivened by concrete, situated practices – and thus we need to take account of the actual *oeuvres* of making publics.

So far, the ways in which urban public space acts as a system to (re)produce ideologies, meanings and practices in Chinese cities is yet to be systematically understood and theorised. Is there a public/private divide in modern Chinese cities? How has publicness been lived and practised in varying geographical and historical contexts? How does public space set limits on, but also enable, collective identities and actions? These questions beg closer readings of the myriad processes in which public spaces are imagined, ideologically coded, performed and practised.[2]

Mindful of the radical incongruence between Eurocentric roots of the treatise on public space and the practices of publicness embedded in social realities of urban China, this chapter outlines two critical perspectives which have implications for a broader research agenda on public space in China. First, I argues that to view the urban public as separate from, and even antithetical to the state, is less tenable in a Chinese context. Instead of an arena of civic agency *external* to the state (in the tenor of the Habermasian conception of the *Öffentlichkeit*), public space in China has always been susceptible to state rationales and agendas of politico-social elites. This chapter therefore sides with Huang's (1993) argument that what exists in China is a 'third realm' lying between the state and society, the construction of which both parties partake in. As the chapter reveals later, it is not uncommon for public space in urban China to be mobilised by the state to advance purposes of governance (for instance, the engineering of 'modernity', 'progress' and 'civilisation', the nurturing of political allegiance, the maintenance of social order and wellbeing etc.) and governmentality (the cultivation of new socio-political subjectivities and values). In this vein, public spaces, even those lived and practised in spontaneous and bottom-up ways, are implicated in the reproduction of state power. It also needs to be acknowledged however that the normative separation of the state and society has never been fully realised even in the West; but that the absence of a Habermasian notion of public sphere makes the entanglement of state and society seemingly natural and far less resisted in Chinese contexts.

Second, despite an acknowledgement that state interventions proactively shape public life and configure the state – society interface, it must be stressed that China is far from bereft of spaces of genuine grassroots agency. In most cases, urban public spaces are not (re)produced through pre-existing civil society as such, but are small, fragmented and transient urban *publics* (in the spirit of Watson 2006), spaces which are unifying and expressive, bringing identities, interests and concerns out of the confinement of private households to be negotiated in common discursive and material spaces (see Chapter 6, this volume).

To engage with these arguments, in this chapter I discuss three scenarios of public space and public life in modern China. I draw on published works in both Chinese and English languages to frame my arguments. While the scenarios I portray are by no means exhaustive of the (re)production of public space in modern Chinese cities, it is my hope that they provide pertinent snapshots of entanglements of historical transition, emerging modalities of state governance and grassroots agency. First I explore the interests of the state and elites in building public parks based on Western conceptions of public space in the late Qing and Republican eras, with the hope of cultivating citizens with 'civilised' and 'healthy' lifestyles. Second, I investigate spatialities of expression and agency, which are not characterised by overt confrontation between the state and society, but by the ongoing interplay between the two realms. Special emphasis is given to densities of social life and interaction and mundane socialities taking place within an intermediary zone between private households in order to investigate broader structures of inequalities and power relations, which give rise to nascent and opaque, yet motivating and inspiring identities of 'public man'.[3] And third, I consider controversies around 'square dancing' (*guangchang wu* 廣場舞) which has increased in popularity and received persistent media interest over the past few decades (see Chapter 14, this volume). In doing so, I argue that public space allows urbanites in post-reform China to negotiate rapid social change as an embodied and performative practice that involves a constant unfolding of pleasures, social relations and subjectivities

Sketching a historical background

In this section, I briefly outline the historical background of public space and public life in urban China. While accurate periodisation is a difficult challenge, I approach fuzzy historical timescales by outlining three phases:

- Ancient Chinese cities (up to the late nineteenth century): as early as the Song Dynasty (circa the eleventh century). After the night curfew system in walled residential wards was lifted, bustling and vibrant street life soon began to emerge in major Chinese cities. This was evidenced vividly by the renowned artwork *Along the River during the Ching Ming Festival* (*Qing Ming Shang He Tu* 清明上河圖). Streets and a kaleidoscope of communal spaces and commercial establishments were not only venues of leisure and entertainment, but also of communal organisations that dealt with collective concerns and welfare. These repertoires of street culture had enduring imprints on civic life in Chinese cities. As Friedmann (2005) observes, in the later imperial era, a well-organised civil society, taking root largely in small communal and public spaces was on the rise and ready to mediate community issues and shoulder responsibilities for various forms of social welfare.

- Late Qing Dynasty and the Republican Era (late nineteenth century – 1949): the later years of the Qing Dynasty were a time when Western imperial influences became palpable in every aspect of urban life in China. As a result, political and social elites were exposed to knowledges of Western urban planning, and recognition that there was a realm beyond private households and neighbourhoods that needed to be collectively maintained, managed and optimised by the state and society, so that the wellbeing of urban inhabitants could be maximised. Concurrently, functional differentiation and rational planning were regarded as the *raison d'être* of progressive urbanism, and cities were being (re) planned and (re)built by following templates of Western metropolises. Emerging spaces included cinemas, cafés, department stores and night salons which contributed to modern sensibilities and new horizons of consumption-led experiences (see Chapter 8, this volume). Shanghai at that time was the archetypical city that embodied a nascent cosmopolitan modernity (Lee 1999), due to the existence of colonial authorities and the influx of foreign and domestic capital. Such new urban development not only created 'a world of strangers' leading to new public orders and moralities, but also sensuous encounters with the phantasmagoria of urban realities (Zhang 2010; also see Chapter 5, this volume).
- The era of the People's Republic (1949 –) : the socialist state approached urban modernity in radically different ways from erstwhile elites. Consumption and leisure activities premised on capitalist modernity were deemed inappropriate for the proletariat culture. Instead, public space during the Maoist time (1949 – 1976) was dominated by monumental structures, mass mobilisation, and political propaganda (Wu 2005; Gaubatz 2008; Hung 2013). The design of public space was not humane or conducive to spontaneous and organic public life, emphasising sublime symbolisms and grand scales. However, in the era of economic reform (1979 – present) economic liberalisation and China's increasing (re)integration with the global economy ensured that to a large degree public spaces were de-politicised, seen as leisure and consumption spaces by policy makers, and that grassroots agency had once again begun to make a presence in the public sphere (albeit to limited and varying extents). Gaubatz (2008: 72) summarises five processes reshaping public space in post-reform China:

> (1) The opening of new or redeveloped spaces to the public through the removal of walls and other barriers; (2) the changing form and function of open squares and plazas in Chinese cities; (3) the commercialisation of public space in retail centres, (4) the emergence of new activities and spaces in parks and art and entertainment venues, and (5) the ephemeral spaces and activities that have become characteristic of transitional China.

Three scenarios of public space in modern China

Cultivating modern citizenry through public space

China's urban modernity, which emerged in the late nineteenth century, was characterised foremost by a socio-political movement orchestrated by the state and elites, under the name of 'municipal government' (*shizheng* 市政運動) (Zhang 2006; Shi 2016). The *shizheng* movement recognised the need for a public sector that provided and regulated urban spaces and services. A central tenet of this movement was that the city, if planned, built and managed properly, would expedite progressive social changes, and produce a superior form of citizenship as well as a sense of belonging to the 'collective good'. This conviction became particularly powerful in the aftermath of the fall of the Qing monarchy in 1911, as the political and social elites aspired to embrace the identity of citizens of a modern, democratic republic. The city, therefore, became a focus of active state intervention – in the view of reform-minded politicians and elites, with physical spaces of the city playing an abiding role in engineering the values, orders and norms of a 'new society'. Another rationale of the *shizheng* movement was to create a democratic realm to which the urban citizenry had equal access – to 'give the city back to the people', especially after monarchical rule came to its demise.

One of the social improvement projects that the *shizheng* movement accomplished was the building of modern public parks. The concept of the 'Western' public park became appealing to reform-minded intellectuals and elites in China as the latter held the strong conviction that the reason for China's subordination vis-à-vis Western imperial power lay in a populace living in incivility, irrationality and ignorance. In this context, public parks were invested with the hope of cultivating modern citizens who led civilised and healthy lifestyles, had a sense of public interests, and observed public order (Shi 1998). Tianjin Park, completed in 1907 in Tianjin, was the first public park project spearheaded by local state and native elites. Soon, major cities such as Beijing, Guangzhou, Shanghai, Chengdu and Wuhan saw a proliferation of state- or elite-funded public parks.

Rationales for the state and elites to advocate the development of public parks centred largely on the cultivation of a modern 'civilisation' (*wenming* 文明). Put in another way, public parks stirred up the modern sensibilities of order, reason and progress. In this vein, to civilise and moralise the people was an idiosyncratic, yet indispensable function that parks in modern China came to bear (M. Li 2009). As Edward Waite Thwing, a Presbyterian missionary based in Tianjin, remarked in 1910, public parks were capable of improving public hygiene and inspiring the wisdom and virtue of the people (Cui 2009). This view was likely to be resonant with the mainstream of Chinese intellectuals and elites.

Public parks were expected to be conducive to public hygiene and health. In fact, public parks were among the first social experiments in China that combined urban spaces with Western notions of public health. This agenda emerged in tandem with a prevalent thesis at that time, which maintained that strong physiques of the people had a direct bearing on a strong nation, and physical health contributed to mental and moral health. As a result, not only were public parks painstakingly landscaped with greeneries to act as the 'lungs' of the city, but also equipped with a diverse array of sports facilities. Visitors were encouraged to engage in bodily exercise and sports (He 2011; Feng 2016). Unsurprisingly, to 'civilise' people from top-down by imposing ideas abstracted from experiences in the imagined 'West' was occasionally met with disappointment. For example, as He (2011) documented, spitting was still commonly sighted in Beijing's parks in spite of the explicit prohibition enacted by the authorities.

Concomitantly, public parks in early modern China incorporated an explicitly *pedagogical* dimension. In the view of the state and social reformers, parks provided space in which educational materials could be stored and displayed, and educational activities carried out. The overriding expectation was that knowledge of the people could be enhanced. Public parks in Beijing, for example, usually had a gauntlet of affiliated facilities such as museums, libraries, exhibition rooms and lecturing rooms (Zhou 2008); educational activities hosted in parks included exhibitions on public health and moral virtues, films, arts training, sports classes, among others (Pang 2015). W. Wang's (2008) work offers a detailed list of the exhibitions taking place in Beijing's Central Park (中央公園) in 1934, the majority of which were themed on Chinese and Western arts, as well focusing on Chinese cultural relics. The programme used by the Education Gallery in Chengdu's Shaocheng Park (少城公園) highlighted a more expansive spectrum of themes including civic virtues, national defence, modern science, children's education, and even the use of electricity (D. Y. Li 2009). Public parks in China were also precursors to bringing the Chinese society into encounter with Western natural sciences, given that many included zoos (Chen 2004).

The development of public parks was underpinned by belief that the everyday lives of Chinese people needed to be radically reformed, to become more civilised and healthy. In this discursive formulation, public parks opened up a horizon of new, healthier urban experiences that distanced people from the vulgar habits of 'gourmandising, drinking, whoring and gambling' (*chi he piao du*, 吃喝嫖賭) (Zhou 2008; D.Y. Li 2009). The state and elites also made great efforts in regulating the demeanours and conducts of park visitors. In Tianjin, for example, a municipal 'Park Visiting Convention' was put in place as early as the 1900s. This protocol specified the rules and orders which park visitors should abide by, including the codes of proper dressing and behaviours (Chen 2004).

Meanwhile, the state and elites not only invested in new public space and public life, but also attempted to regulate already existing and spontaneous forms of lived spatial practice. Traditional communal spaces such as the teahouses, which had hitherto enjoyed considerable autonomy from the state, were increasingly subject to state intervention on the grounds of hygiene, civility and order (Wu 2009; Chen 2008). Being taken into the nomenclature of state discourses and actions often fundamentally redefined and remade grassroots social life, although the latter was by no means hijacked by state rationales. For example, the urban historian Wang

Di's work on street culture and teahouses in Chengdu is enlightening here: while the streets and teahouses were quintessential to a sense of collective belonging and the relative autonomy of the grassroots society, accommodating leisure, commerce, socialisation and even the arbitration of community conflicts, they were also situated in tensioned relationships with social reformers/elites, who derogated streets and teahouses as the hotbeds of indolence, gossip, incivility and disorder. A plethora of reforming agendas was therefore enforced by the police and other authorities, aiming to regulate the forms and contents of popular entertainments and socialities (D. Wang 1998, 2001, 2003, 2008).

Spaces of expression and grassroots agency

One important scenario that constituted imaginations of Western public space was notions of civic autonomy, agency and unfettered expression. Despite the idea of modern public sphere being alien to Chinese intellectual, cultural and political traditions, notions of public space in modern China nonetheless occasionally represented expression of material and embodiment of grassroots agency. Nonetheless, in lieu of a deliberate separation of the state and society, there was more often than not an intricate intersection of the two realms in the making of urban public space in China. In tandem with this hybrid nature, the use of public space was an expression and an activity that both the state and grassroots actors adopted.

As I have argued elsewhere (Qian 2014a), akin to civil society, the state also seized the advantage offered by emerging discourses surrounding public space to propagate dominant values, in order to socialise and indoctrinate the masses. To elaborate this point, let me go back briefly to the public parks in modern China. For the state, the public park was viewed as an ideal medium for disseminating political ideologies and nurturing national identities. In the Republican era, public parks were already heavily used for assemblies of political propaganda. Moreover, many parks were dotted with sites and landmarks engraved with political and nationalist semiotics (Chen 2004). The Kuomintang government, for example, was notably keen to assert the legitimacy of its revolutionary campaign by building memorial landscapes and spaces. Intriguingly, the Republican era, albeit short-lived (1911 – 1949), witnessed the construction of over 140 parks named after Sun Yat-sen, the founding father of this regime. Concurrently, however, public parks also acted as settings for political rallies spontaneously organised by the civil society (Zhou 2008; W. Wang 2008). Also, in Chengdu, parks were the seats of the activisms of various trade unions formed by grassroots workers (D.Y. Li 2009). In general, at the interface of state and society, how the voices and claims of one realm outweighed, or converged with those of the other was far from definite, and was contingent of course on the turbulent and highly volatile socio-political situations of China in the early twentieth century.

Such arguments can also be applied to spaces such as teahouses (茶館) and opera houses (戲園). In Shanghai's traditional opera houses, for example, progressive actors developed new pieces of opera that contained explicit political claims and discontent (Fang 2006). Moreover, in teahouses in Chengdu, vernacular spaces of communal socialities, boundaries between gossips and serious talks were unclear. Teahouses thus became politicised, fraught with not only impromptu exchanges of political discourse, but also overt political actions. As Wang Di's (2008) book richly documents, in the early years of the Republican era – a time characterised by relentless efforts of state building and endless political turmoil – teahouses were inextricably enmeshed in broader political struggles. Not only were the employees of teahouses involved in the protests against state restrictions and tax increases, but the teahouses themselves also became central to the communication and exchange of information on activisms and struggles outside the relative comfort and cosiness bestowed by tea drinking. Outbursts of complaints on political uncertainty, rapacious state power and corrupt government were common in teahouses. Indeed, later during the Anti-Japanese War (1937 – 1945) the state exploited the opportunities offered by socialities of teahouses for the purpose of political mobilisation. While the state and elites had historically criticised teahouses for perpetuating lives of leisure and insensitivity to the destiny of the nation, they now attempted to insert patriotic and anti-Japanese pedagogies into popular entertainments. Accordingly, 'model' teahouses were selected as key nodes in a state-led network of political propaganda.

FIGURE 2.1 Teahouse in Chengdu

Source: Junxi Qian

Investigating experiences of the early years of urban modernity in Chinese cities thus help to elucidate my earlier contention that urban spaces in China were incubators of grassroots urban publics, which were not as universal or homogenous as Habermas's thesis suggested, but fragmented and crosscut by concrete identities and differences based variably on occupation, communal ties, kinship, places of origin, among others, and might overlap, ally, or compete with each other (Qian 2014b). As Rowe (1990) suggests, this notion of publicness is particularly relevant when we look at the proprietorship of collective goods and the management of social services in China, which necessarily entailed the expansion of popular sovereignty and participation.

Indeed, I want to draw on Rowe's (1984, 1989) works on late Qing Dynasty Hankow (漢口) to shed light on the fluid, transient and practice-based nature of these urban publics. In the first of his two books on commerce, society and community in Hankow – a burgeoning commercial hub in central China – Rowe (1984) proposed the that various professional guilds in Hankow had become the de facto makers of rules governing social and economic practices in the city, bypassing the state and creating autonomy for a maturing civil society. He later noted that communal organisations and guilds even went into partnership with the state in the provision of welfare, and civil society initiatives played an important role in the construction of civic infrastructure (Rowe 1989). In this vein, urban spaces – poorhouses, orphanages, benevolent halls, streets, bridges, piers, flood relief systems, and even cultural and leisure facilities – were not only physical expressions of the organising power of grassroots urban publics, but also the spatial foci around which the consciousness of a 'public man' was constituted and sustained. As such, grassroots publics were effective in mitigating relationships between the state and society, and between various sections of the civil society by mediating group contestation and quenching protests; they were even involved in devising impromptu arrangements to enforce order and security. To sum up, the urbanism of late Qing Hankow cannot be conceptualised without sufficient attention to a vibrant 'third realm' that navigates between the state and private urban dwellers (for a comparable case of Beijing in the Republican era, see Strand 1989).

It is broadly agreed that long-standing and emerging notions and practices of urban public space were largely undermined after the founding of the People's Republic in 1949. This view might be more or less true to the Maoist era, due to only rare documentation of bottom-up public life in the Maoist era existing to compare with previous generations. However, in the aftermath of 1979, the economic reform and the loosening of state control created opportunities for the revival of grassroots agency, and the implication for public space has been profound and remarkable. Indeed, when I first arrived at Edinburgh University to start my PhD study, I was reminded by my supervisors that China in fact had the most renowned public space in the world, Tiananmen Square, which had been consecrated as a totem of resistant politics (Hershkovitz 1993). Yet, grassroots agency is not restricted to spectacular movements; also, a symbolically laden space alone is not politically productive in its own right, but more analytical weight needs to be given to the mutual constitution of space and situated practices and performances.

In line with these points of view, Xing's (2011) study of the Workers' Cultural Palace in Zhengzhou and my own work on Red Song singing in Guangzhou (Qian 2014b) both focus on the formation of political publics by way of collective *cultural* activities. The Workers' Cultural Palace, an emblematic space of Maoist mass mobilisation, witnessed the retreat of the state in organising leisure and cultural activities in the reform era. This left the workers to (re)construct working class subjectivities by acting from below. While singing, dancing and presenting impromptu theatrical performances were ultimately for the ends of fitness and exercise, the workers also appropriated Maoist mass culture and slogans to insinuate severe social inequalities and worker disempowerment in the post-reform era. Translating Maoist nostalgia into concrete political actions, the workers deliberately politicised public space by commemorating Mao and distributing leaflets with anti-market, anti-reform sentiments (Xing 2011). The blurring between the cultural and political, between leisure and political activism is also evident in Red Song singing in Guangzhou (Qian 2014b). The Red Song was part and parcel of the socialist state's initiatives to win the collective consent and conformity of the people by shaping and controlling habitual cultural experiences. In post-reform China, where socialism no longer finds anchor in most socio-economic realities, collective singing of Red Songs in public parks was practised by singers as an innovative way of articulating political identities (see Figure 2.2). These identities not only drew from the discursive

FIGURE 2.2 Collective red-song singing in Guangzhou

Source: Junxi Qian

contours contained in the lyrics, but also embodied performances and social interactions in the immediate socio-spatial settings of singing. Neither were the identities fixed or monolithic, but constituted by mixed feelings and sentiments – both persisting in allegiance to the party-state and critical reflections on the decline of socialism in the post-reform era.

Square dance and public leisure in post-reform China

Public space in post-reform China has witnessed a proliferation of spontaneously organised leisure, most often including the form of public dancing in groups of variable sizes (see Chapter 14, this volume). Standing in lines, dance takes place in unison to music from portable stereos. The types of dances includes Chinese folk dances, ballroom dancing, Latin and Tango, hip-hop and street dances. Participants consist mainly of middle- and older-aged urban residents, but younger generations are also visible. Public dance is clearly a source of great rejoicing for the dancers, who participate not only for exercise and to see improved health, but also to revel in performing 'the self' and collective identities (Orum *et al.* 2009; Qian 2014c; Lin and Bao 2016). In China, public dance of this nature is dubbed 'square dance' (*guangchang wu* 廣場舞), with media depictions often focused on describing manifestation of a revival of organic urban socialities owing to the demise of Maoist social control, or a public nuisance that encroaches on the 'collective good' by occupying spaces and being too noisy (see Chapter15, this volume).

Theoretical engagement with square dancing has mainly pursued two distinct lines of inquiry to explain the spread of spontaneous public leisure, now scattered in almost all urban public spaces from grandiose squares to narrow roadside locations. The first approach has been concerned with the ways in which public leisure is mobilised by urban inhabitants to negotiate restless social changes, construct collective belonging, and find anchors of identities, in response to concerns that traditional society is being engulfed by consumerism and individualism. In these formulations, commentators have investigated normative values and symbolic meanings borne out in public leisure, and by revealing correspondence between spatial practices and social changes. Jayne and Leung (2014), for example, suggest that public dance can be read as rediscovery of collectivism and as a playful act, a response to the regulation, ordering, and rationalisation of urban spaces. This argument is resonant with my own study of public leisure in Guangzhou (Qian 2014c) which highlighted how public dance acts as an antidote to logics of commodification and individualisation in post-reform urban China, and that to a certain degree resurrection of Maoist values of collectivism is clearly discernible in the negotiated conventions and ethics of leisure-based socialities.

In a parallel set of arguments, commentators have pointed to the importance of understanding how public leisure can challenge pre-existing identities and cultural orientations. As such, leisure activities in public spaces can be best elaborated as an *actor network*, a dense field of material and immaterial unfolding and becoming, in which subjectivities and the spatialised self are refashioned by way of embodied, performative practices and engagements. Public leisure, in this vein, can be analysed from a more-than-representational perspective and conceptualised through embodied, emotional and affective 'ways of being in the world' (Jayne and Leung 2014). This approach is not antithetical to the first with both representation and discursively formulated meanings contributing to instantaneous moments of movements, emotions and affect. Analysis of these improvised social dramas, therefore, needs to be centred on the bodily choreographies, intimate social interactions, and the socialisation of bodily comfort and discomfort (Jayne and Leung 2014). Indeed, I have argued that identities and meanings that hold together transient collectives, who are willing to invest large quantities of money and labour into organising leisure activities, emerge from situated encounters and social interactions, and performative presentation of the self (Qian 2014c). Richaud (forthcoming) also develops this argument by examining old-aged leisure participants in Beijing's parks. Richaud highlights the importance of activity-based friendship networks among park users which emerges from ludic encounters and situated interactions, yet without superseding the modern urban ethics of anonymity, indifference and separateness. Richaud goes on to suggest that, to avoid reifying broader social transformation as something that mechanically makes actors behave, analysis should be focused more on the *present*, the situations themselves.

In summary, academic engagement with public dance points not to an urban utopia free from various axes of social inequalities and antagonistic social relations. Qian (2014c), for example, points to orthodoxies regarding 'correct' ways of organising activities and presenting the self which vary greatly between social classes (echoed by Jayne and Leung 2014). For example, middle class participants tend to segregate from other groups and disparage rural migrants' activities as vulgar, uncultured and of low quality (see Chapter's 10 and 13, this volume). Meanwhile, in contrast to English-language publications, which are more likely to eulogise public leisure as epitomising the 'alive and well' public man in the Chinese city (Orum *et al.* 2009), Chinese-language works tend to be cautious of the fact that public leisure can be a challenge for urban governance. On the one hand, loud music broadcast for public dancing often leads to tension between 'public' activities in noisy cities which impact on 'private' domestic lives and which sometimes has escalated in open, violent conflicts (see Chapter's 4 and 15, this volume). On the other hand, such mass cultural activity is occasionally depicted as being reminiscent of catastrophic enforced 'collectivity' pursued during the Cultural Revolution, which, in the view of many, is at odds with discourses of privacy and free choice in the post-reform context (Dai 2015). Theoretical debate thus continues regarding the extent to which mass leisure in urban public spaces should be interpreted as a testimony to the 'thrown-together-ness' of the city as unconstrained encounter (Amin 2008), or as an erosion of public space by selfish interests. Similarly, there is much to be gained by theoretical and empirical interrogation of 'blurring' of boundaries between the private and public, between collective memories and new social zeitgeists, between communal belonging and depictions of encounters between strangers in contemporary cities (see Chapters 5 and 7, this volume). These topics amongst numerous others highlight important avenues for further research and the fruitful terrain offered by critical engagement with public space in Chinese cities.

Conclusion

In a recent review of progress He and Qian (2016) suggest that one way for Chinese urban studies to realise its potential of contributing to global research agendas is to de-naturalise and destabilise fundamental concepts that continue to dominate Western-centric urban theory. Challenging various theorisations of public space is undoubtedly one area where scholars of Chinese urbanism have much to offer advancement of urban theory. Towards that aim, in this chapter I have pointed to productive moments of critical dialogue and points of departure between Western and Chinese conceptions of public space and urban public realm. Drawing on empirical evidence I have argued, for example, that urban public space in China can be productively viewed as a hybrid realm over which both state and grassroots society strive to exercise influence and control, and that in Chinese cities transient, but yet lively publics can be theorised through situated exigencies, needs and interests.

Indeed, I have argued that akin to public space in the Anglo-European intellectual tradition, public spaces in China and the urban public realm in general, are *social* constructs that various actors construct to respond to and negotiate the tempest of social changes. As Rowe (1984, 1989) demonstrates, myriad urban collectives and communal organisations arose out of the rapid urbanisation and commercialisation set in motion by China's (forced) incorporation into the colonial world system. Similarly, the popularity of square dance in post-reform China highlights how grassroots agency responds to concerns over individualisation, commodification and alienation. Such socio-spatial practices point to the importance of Richaud's (forthcoming) view that active urban public space is *overdetermined*, rather than determined, by social contexts – urban publics are brought into life *in situ*, in the continuous unfolding of practices, actions and meanings. Critical perspectives that take into account historical contingencies and the messiness of concepts and lexicons thus not only have much to offer our understanding of Chinese cities but also have a vital role in advancing international debates regarding urban theory.

Notes

1 In this chapter, public space refers primarily to physical urban spaces, although this necessarily implies the importance of taking seriously (non)human agency.
2 It is often argued that there is a 'fuzziness' between public and private space in Asian cities (e.g. Drummond 2000). However, I argue that there are significant limits to such arguments and a need to interrogate in more sophisticated ways relationalities of 'public' and 'private' space.
3 Throughout the chapter, the term 'public man' is used as shorthand to represent longstanding academic debate. This does not however exclude the importance of theorising 'public women'.

3

URBANISM AS A STATE PROJECT

Lessons from Beijing's Green Belt

Hyun Bang Shin and Yimin Zhao

Introduction

As of 2011, national statistics in China were suggesting that the share of China's urban population had exceeded the 50 per cent threshold for the first time in its history. This is taken as an endorsement of China's entering an 'urban age', though such claims in official urban discourses have been criticised in recent literature (see for example, Brenner and Schmid 2014; see also Shin 2018). Higher urbanisation rates are supported not only by the natural growth of urban population but also by the conversion of rural villagers into urban citizens. The latter process entails land grabbing, which converts existing agricultural farmlands into urban construction lands to accommodate the provision of new real estate properties (e.g. new apartments, offices), infrastructure (e.g. motorways and high-speed rail) and production facilities (e.g. industrial parks). What do all these mean for villagers who lose access to their farmlands?

We often hear the frustrations of villagers whose lands are violently taken away against their will with no or poor compensation (e.g. Hoffman 2014; Johnson 2013; Pomfret 2013). Sargeson (2013) argues that violence is an integral element of China's urbanisation project, authorising urban development. In this chapter, we show that such use of state violence goes hand in hand with another dimension of state action, that is co-option of villagers (cf. Gramsci 1971) by the imposition of what Henri Lefebvre (2003) refers to as 'official urbanism'. Drawing on Lefebvre's critiques of urbanism, this chapter aims to reflect upon the use of official urbanism to advance China's 'urban age', and addresses two analytical objectives by dissecting green belt policy in Beijing. First, we demonstrate China's urbanism as an institution and an ideology is a state project: it is integrated with both economic and political practices, and plays a critical role in sustaining the state strategy of land-based accumulation. Second, we also illustrate that official urbanism, as an ideology, has been successfully instilled into the national ethos, imposing it upon the population (especially villagers) as a new and desirable way of life, which in turn supports the state's project of urbanism. We conclude that urbanism is one and the same expression of politics of urban space, with the Party-state's ideological, economic and political ambitions put at the centre. For this reason, any meaningful approach to critiquing existing sets of urban knowledge and practice that produces urban inequalities and injustice in contemporary China should start from negation of the 'official urbanism'.

Empirically, this chapter focuses on the making and commercialisation of Beijing's green belts; it investigates the ideological, political, and economic mechanisms for erecting the green belts project on the one hand, and, on the other, uncovers the discursive moment when this project was successfully instilled into the ethos of the population as the *only* desirable way of life in urban change. This story hence shows vividly the juncture where two aspects of the 'official urbanism' – institution and ideology – were dialectically articulated with each other.

Urbanism and the state

For Louis Wirth (1938), the city not only refers to larger dwelling places and workshops, but also marks 'the initiating and controlling centre of economic, political, and cultural life' in the era of industrialisation when the role of the city looms large. One methodological implication of these socio-economic changes is that the urban-industrial mode of living rose to such a significant status that it could be juxtaposed with the rural-folk society as two ideal types of communities (ibid. 3). Drawing on this recognition, Wirth then defines urbanism as a way of living (in an industrial society) and calls for the attentions of American sociologists to the urban mode of human association. He rightly asserts that urbanism should not be confused with the city or industrialism: as the new mode of life, urbanism is neither limited by 'the arbitrary boundary line' of the city (ibid. 4) nor solely conditioned or determined by modern capitalism and industrialism (ibid. 7). In efforts 'to discover the forms of social action and organisation' (ibid. 9), Wirth identifies three fundamental attributes of urbanism to direct sociological studies of the city: population size, population density in a settlement, and the heterogeneity of urban dwellers. Yet, there is a marked lack of attention to the politico-economic relations and processes underlying the production of urbanism.

In *The Urban Revolution*, Lefebvre (2003: 6) argues that urbanism has often been understood as 'a social practice that is fundamentally scientific and technical', while in reality it 'exists as a policy (having institutional and ideological components)'. For Lefebvre, urbanism is indeed 'a form of class urbanism' (ibid. 157), and '[i]t is only from an ideological and institutional point of view ... that urbanism reveals to critical analysis the illusions that it harbours and that foster its implementation' (ibid. 164). Indeed, urbanism is a superstructure of the society (ibid. 163), one that is ontologically connected with both the logic of capital and the rationale of the state. While Lefebvre's discussions of urbanism are rooted in his critique of capitalist societies, it has much to enlighten researchers on urban China where the emphasis on urbanisation has been stressed heavily by the state.

Facing the onset of intensifying urbanisation during the early years of market reform, urbanism has been subject to scholarly attention among China researchers. Ma (2002) is among the first authors who adopted the concept of urbanism in setting research agendas for the study of China's urban transformation, referring to urbanism as 'the *nature* of urban life and of cities as places that are seen as impregnated with geographic, social, economic, cultural, political, and ideological meanings' (ibid. 1556; original emphasis). He further develops his perspectives on urbanism in his subsequent work, in which he provides a summary review of historically identifiable forms of urbanism (Ma 2009). Here, instead of focusing on the connotation of urbanism as a way of life in the industrial age (cf. Wirth 1938), Ma broadens its scope to such an extent that China's 5,000-year history is divided into five periods and narrated using urbanism as an anchor. While Ma (2002: 1563) highlights the importance of investigating 'the central role that the Party-state has played in affecting the processes and outcomes of urbanization and urbanism', how the working of the Party-state and the particular configurations of the political economy of reformist China produces its own urbanism is left unanswered and remains as a challenge for other scholars to address.

A number of China scholars have paid attention to the need of understanding urbanism as an embodiment of the dynamics of the state, space and social fabrics. Cartier (2002), for instance, resorts to transnational urbanism to examine Shenzhen's attempt to transform into a world city by producing a new city centre realised through the state-dominated enterprise that makes use of 'plans, ideologies and representations of domestic and transnational élites to establish legitimacy' (ibid. 1513). In Cartier's analysis, the spatial and politico-economic processes of producing urban landscapes are uncovered, which in turn enable her discussions on how 'trans-boundary and transnational spheres of economic activity and cultural forms' (ibid. 1518) are articulated to endorse the 'spiritual civilisation campaign' of the state in the urbanisation process.

Hsing (2010) also analyses the nature of urbanism in China's urban transformation, seeing it as a set of discourses that legitimise the state's conduct. For example, the idea of 'new urbanism' is linked to the idea of modernisation in endorsing the state's consolidation of its 'opportunity space' for producing and exploiting the urban space, where the land businesses of the state are developed to an unprecedented scale. Hsing (2010: 54) makes a similar argument through the comparison between development zones in the 1990s (as a symptom of industrialism) and the new city projects in the 2000s (which signal the rise of urbanism), concluding that the term 'urbanism seems to have provided a unifying ideology for the political elite'. With urbanism as a shared analytical concept, Cartier registers local politico-economic mechanisms in producing the urban space, while Chen and Hsing both show how urbanism has been deployed by the Party-state in China as a core ideology.

The recognition of urbanism as both an ideology and an institution has also been developed in recent discussions on eco-/green urbanism and speculative urbanism. Hoffman (2011) analyses the making of a 'model garden city' in Dalian and uncovers urban modelling – such as 'green urbanism' – as a governmental practice. This practice not only shapes and produces the urban space with ideas from elsewhere, through the mechanism of inter-referencing of policy discourses, but also remakes urban subjects through the combination of greening practices with 'the fostering of civilised and quality citizens' (ibid. 67). Pow and Neo (2015: 132) draw on Hoffman's conclusion and explores the case of Tianjin Eco-city to decipher the 'new forms of ecological urban imagineering and socio-ecological life-worlds'. They conclude that Tianjin Eco-city is at best an 'ecological imagineering of green urbanism' (ibid. 139), thus serving as a vivid case of the complex interactions between urban sustainability, urban entrepreneurialism and neoliberal urbanism. Also focusing on the Tianjin Eco-city, Caprotti (2014; Caprotti *et al.* 2015) investigates the concrete connections among environment discourses, the market logic, the social fabric of dispossession, and the rise of the new urban poor at the juncture of 'green capitalism'. By asking whom the suffix 'eco' is for, they uncover politico-economic mechanisms and social effects of the Chinese agenda of the 'ecological modernisation', which only result in the booming of elite urbanism and 'the construction of eco-enclaves' (Caprotti *et al.* 2015: 509), consequentially sacrificing the ordinary citizens and their everyday life.

Elsewhere outside the scope of China urban studies, urbanism as both an institution and an ideology also underlies the work of Goldman (2011) who conceptualises 'speculative urbanism' in his study of peripheral urbanisation of Bangalore, India. Drawing on the epistemological critiques of urbanism, Goldman sees land speculation and dispossession of people at the urban periphery as the principal state business in Bangalore's making of the 'next world city' (ibid. 555). Such practices institutionalise the 'temporary state of exception' into the 'new forms of "speculative" government, economy, urbanism and citizenship' (ibid. 555), which defines what he labels 'speculative urbanism'. This discussion is noteworthy for its refusal of the use of urbanism only at its face value. Instead, urbanism is seen as the critical moment in the urban process, which is shaping, and being shaped by, socio-historical conditions and politico-economic mechanisms constituting urban changes. With this concern, Shin's (2012) work on China also recognises the preponderance of place-specific accumulation strategies of the Party-state through the examination of mega events as urban spectacles. These strategies are deployed in a geographically uneven manner to shape the Chinese version of 'speculative urbanism' (Shin 2013: 2014).

In this chapter, we highlight how urbanism works as a concrete mechanism through which the urban space is utilised by the state for its own goals – no matter how dynamic and transient these goals are – so as to sustain its legitimacy and to reproduce social/power relations. Put it in another way, we see urbanism as a permanence of the daily life under urbanisation where politico-economic concerns of the state are crystallised into a coherent ideology and embodied within the associated institutions. Through the examination of the rise of China's official urbanism in the green belts project, we argue for greater responsibility of researchers to investigate these tangible institutions and ideologies rather than to adopt urbanism only as an empty signifier that obscures the nature, agency and rationale of the state.

Official urbanism as a state project

In his critiques of the potential urban strategies in the socialist countries, Lefebvre (2003: 147) presents an explicit definition of 'official urbanism'. On the one hand, socialist urbanism is also a type of urbanism, and is hence 'not very dissimilar from capitalist urbanism' (ibid. 147) in terms of its nature as an institution and an ideology. On the other hand, because of an overwhelming emphasis on the ideology of industrial production, the socialist version of urbanism tends to exhibit such characters as 'less emphasis on the centrality of exchange ... greater access to the soil ... attention to an increase in the amount of green space ... and the zero degree of urban reality' (ibid. 147). His observation was made in 1970, when socialism as a state institution and a seat for geo-political power had not yet disintegrated, and when the Stalinist model of economic growth was still predominant in most socialist countries (including China). The ideology of industrial production and 'the zero degree of urban reality' rendered China's urban space into the situation labelled by Ma (1976) 'anti-urbanism'. This condition in turn laid a politico-economic foundation for importing the idea of 'green belt' into Beijing out of its concern for green space (Zhao 2016), and, accordingly, shaping the 'official urbanism'.

Simply put, the idea of green belt was imported to Beijing from Britain and the former Union of Soviet Socialist Republics (hereafter USSR) in the 1950s, which was seen at that time as a promising ecological goal of its socialist transition (Beijing Archives 1958). Indeed, it arrived in Beijing at the height of a socialist-utopian campaign named the 'Great Leap Forward' (大跃进, *dayuejin*). This was a period when the Chinese people were mobilised by Mao Zedong to 'surpass Great Britain and then catch up with the United States' (超英赶美, *chaoying ganmei*). Among targets of this campaign, 'gardening the earth' (大地园林化, *dadi yuanlinhua*) was set as a socio-ecological goal (Chen 1996; CPC 1958). Six decades have passed since the green belt arrived in Beijing as an idea, and a sea change was witnessed with regard to China's social and politico-economic conditions (Ma 2002; Wu 1997). Nevertheless, the green belt is still defining Beijing's urban master plan right now, as it did in 1958 when the very first modern master plan of Beijing was drawn out (Yang 2009; Zhao 2016). Between the survival of the green belt in the master plan and the changing urban political economy lies the secret of the Party-state and the role of its official urbanism.

To understand changing rationales of the Party-state hidden behind official urbanism, it becomes necessary not only to investigate the institutional dynamics but also to explore how far they are achieved spatially in the urbanisation process. The concrete mechanism through which urban space is utilised by the state for its dynamic goals is the milieu in which we can recognise and define the official urbanism. In this section, we focus on three aspects of such a mechanism that shapes Beijing's green belt, namely, the ideological, the political and the economic.[1]

Shifting ideological connotations

The articulation between Beijing's green belt and official urbanism was established through the ideological connotations of the former. In the Maoist era, official urbanism was set in tune with the ideology of industrial production. The imposition of the green belt at that time was mainly serving the ambition of industrialisation-cum-modernisation by segregating industrial areas from residential ones on the one hand (BMCUP 1987: 199 – 200) and by 'gardening the earth' on the other (Chen 1996). In the 1958 Beijing Master Plan, the Beijing Municipal Government (hereafter BMG) claims (BMCUP 1987: 206 – 207) that:

> In the last few years, the scale of redeveloping and expanding the city proper has been huge. The urban layout should not be too concentrated, and a dispersed and clustering model is to be applied from now on. There should be green spaces between clusters: 40% of the city proper and 60% of inner suburbs are to be greened. In the green spaces, we will have woods, fruit trees, flowers, lakes and crops.

This master plan was drafted with direct instructions from the USSR experts, who arrived at Beijing in April 1955, bringing the Soviet version of modernist planning principles together with them (BMCUP 1987: 32). In this plan, the form of the imagined green belt largely followed the modernist planning canon (such as zoning techniques and the landscape of the garden city), while its content was defined by the utopian vision of making China modern and compromised by the industrialisation of the city. Such a utopian experiment is transient and the content of the belt is accordingly transformed together with the political economy. But first of all, it is the ideological connotations of the green belt that change before anything else.

In the post-Mao era, the previous focus on industrial production in China's official urbanism gave way to such issues as social order, hygiene and internationalisation. In 1980, the Central Secretariat of the Communist Party of China (hereafter CPC) issued an 'important instruction' to the BMG, which highlights that 'Beijing is the political centre of our country, and it is also the centre for international contacts … Economic development policies should also be changed according to the nature of the capital, and heavy industries should cease to be developed' (BMCUP 1987: 75). In the 1982 Master Plan, it was further stated that Beijing was to be the 'political and cultural centre of our country' (BMCUP 1987: 78). Ten years later, this was revised again to refer to Beijing as the 'modern and international city' (BMCUP 1992; Zhang 2001: 274 – 275). Hence, the Maoist (industrial) imagination of 'being modern' was transformed into the yearning to be 'international', and this in turn restructures both the official urbanism and the role of the green belts in line with the new urban development direction.

One of the key methods to put the yearning for internationalisation into practice was to host such mega events as Olympic Games. Jia Qinglin, the then Party Secretary of Beijing, claimed in an official meeting that bidding for the 2008

Olympic Games was not only 'a historical opportunity to accelerate Beijing's development in the new century and to move Beijing forward to a modern and international metropolis', but also 'a perfect approach to show our achievements in the city's modernisation process and to augment its international reputation' (*Beijing Youth Daily* 2000). The role of green belts loomed large here. In Beijing's first bid for the Olympic Games in 1993, its level of pollution had dissatisfied inspectors from the IOC, indeed partly explaining that failure (China Internet Information Centre 2001).[2] As the BMG noticed the importance of the environmental issue in international assessments, represented by IOC's preferences and inscribed in the 1996 edition of the Olympic Charter (IOC 1996), every effort was at once made in Beijing to meet these expectations – for example, it promised in 2001 that a green belt of more than 100 million square metres would soon surround Beijing (BOCOG 2001).

The BMG stressed the role of ecology and environment to such an extent that 'Green Olympics' was ranked the most important initiative (Green Olympics, Sci-Tech Olympics, and Humanistic Olympics) in their bidding for the 2008 Olympic Games (Xinhua News Agency 2001). And through this initiative, green belts were brought to the fore once again in reshaping this city. The concern of green and the appeal to modern-international were juxtaposed here, revealing the new focus of official urbanism and setting up a substantial and legitimised pretext for completely different political and economic ambitions of the state.

Political mobilisation through institutional restructuring

On 29 September 1999, five months after submitting its formal application report to the IOC, the BMG held an Office Meeting for Mayor and Deputy Mayors and decided to establish a new agent entitled the 'Beijing Leading Group for Constructing the Green Belt Area' (北京市绿化隔离地区建设领导小组, *Beijingshi lvhua geli diqu jianshe lingdao xiaozu*; hereafter BLGCGB) (BMG 1999). This leading group was designated as the municipal agent taking charge of all issues related to the green belts project. On 2 March 2000, the BLGCGB held its first formal meeting, headed by the then Mayor Liu Qi (BMG 2000a), to establish its General Headquarters (总指挥部, *zong zhihuibu*). This meeting also witnessed the setting up of a goal by Mayor Liu to finish the greening of 60 square kilometres within the next three years, considerably quicker than the ten years previously planned. Here, 'a fierce battle [was] needed immediately', in the ensuing nine months, 'to meet the annual goal of greening 20 square kilometres' (ibid.). Furthermore, since it was a huge challenge, Mayor Liu continued, 'our cadres are encouraged to break through traditional doctrines and regulations and figure out special measures for this special task; and with these measures new institutions can also be erected' (ibid.).

Dozens of documents were released by the BMG in the next couple of years to institutionalise and systematise their ambitions oriented around the green belts. The very first document as such was a scalar one to get rid of the central government's ban on occupying arable land for urban constructions. In May 2000, the Ministry of Land and Resources of China established a 'Coordination Liaison Group for Constructing Beijing's Green Belt' and issued a document entitled 'Instructions on the implementation of Beijing's green belt project' (Chai 2002: 6). Its special institutional arrangement was to lift the central ban on the BMG by setting up two exceptions. First, the use of arable land in the green belt for planting trees could be registered as an internal adjustment to the agricultural structure, hence was exempted and permitted. Second, occupying arable land to construct resettlement housing for local villagers could also be allowed, insofar as the original settlement was to be demolished and greened (ibid. 6 – 7). Such exceptions were formalised in a local ordinance issued by the BMG (2001a), adopted as a new institutional foundation for directing the state actions to lead the green belts project.

After the exceptional revision of the state's land policies in the green belt area, the BMG set up a new and distinct governing mode for the green belt area. The two principles underlying this mode were called 'special issues, special treatments' (特事特办, *teshi teban*) and the 'all-in-one service package' (一条龙服务, *yitiao-long fuwu*) (BMG 2000b; BLGCGB 2000a, 2000b). Because 'it was challenging to achieve the goal of greening 60 square kilometres in only three years' (BMG 2000a), as the BMG claimed, the previous urban plan was to be adjusted properly to fully respect opinions in the localities so as to accelerate the greening process (BMG 2000b). On top of this, the related land, housing, and administration policies were all subject to revision to clear the way for the green belts project. Since the Olympic bidding rendered the green belts project politically urgent, green light had shone everywhere in the bureaucracy to further accelerate related practices in the exceptional space thus created (BLGCGB 2000a, 2000b; BMG 2001b). For instance, some of the compulsory documents in the approval process were exempted (e.g. feasibility study reports for construction projects) (ibid.) to simplify the procedure and improve efficiency. The political manoeuvres as such, however, did not induce the expected outcome of a green landscape of the city – which only appeared on paper (Yang and Zhou 2007). Such an unexpected result cannot be understood if the state's land businesses were not included in our discussion of the green belts.

Land businesses: the economics of the green belts

It has been widely recognised that the state's monopolisation of urban land and the commercialisation of houses were put at the top of its agenda since the 1990s (Hsing 2010; Lin 2009; Wu 1995, 1997), and the local state has become *de facto* landlords (Shin 2009b). In practice, however, specific land institutions were not established at once. It was only in 1998, after the second revision of the Land Management Act (NPC 1998), when the state underlined its ethos of the use of land resources, which reads: 'we should insist on the simultaneous exploitation and saving of land resources, with saving coming first' (State Council 2000). With the State's Council's encouragement, the BMG set up a 'land reservation system' on 31 January 2002 to consolidate its monopolistic power in the booming land market (BMG 2002a). However, the green belts were excluded from the newly established land reservation system, and thus from the land market. This exclusion should be interpreted through a scalar perspective on the change of land leasing methods at the time in China.

In the 1990s, land leasing in comprehensive development projects was found to be dominated by closed-door negotiation (协议出让, *xieyi churang*) rather than auction (拍卖, *paimai*) (Wu 1995, 1997; Fang and Zhang 2003). Many researchers in China urban studies at the time tended to see such phenomena as a symptom of China's immature land market and to call for more marketisation measures (Ho and Lin 2003; Zhou 2004; Zhu 2000). Nevertheless, the key issue in this regard would not be the maturity of the land market, but territorial-scalar politics and its effects on the everyday practices of state agents. It is gradually made clear that the closed-door negotiation was a rational choice for the local state agents, which enabled them to maintain autonomy in urban space production and also to obtain monetary revenues and hidden benefits for themselves (instead of sharing them with the central government) (Wu 1997: 660; Wu *et al.* 2007: 6 – 8). In the light of the revision of the Land Administration Law in 1998, the State Council issued a series of land regulations in order to change its weak position in the process of land commodification. Two consequent ordinances were released by the Ministry of Land and Resources (hereafter MLR) in 2002 (No.11) and 2004 (No.71). In particular, the State Council aimed at introducing a quota system for land use conversion so that the amount of newly made urban construction land was to be regulated by the central state (Xu and Yeh 2009). The decision was so significant in affecting the land and housing market that the media at the time labelled it China's new 'land revolution' (Hsing 2010: 48; MLR 2002, 2004).

The responses from local governments were anything but obedience to the centre. In Beijing, for example, the BMG issued a local regulation soon after the No.11 Ordinance, listing four *kouzi* (口子; i.e., loopholes for evading central orders) (BMG 2002b). This measure allowed four types of projects to evade the central state regulations: they were projects in green belts, in small towns, for redeveloping old and dilapidated areas, and for developing high-tech industries. In the end, as Hsing (2010: 52 – 53) describes, most urban projects can be categorised as part of these *kouzi* if they obtain an endorsement from municipal or district officials. From June to October 2002, the BMG leased out nearly 90 million square metres of land through closed-door negotiations in just four months, all legitimised under the above four exceptions, while the total area of all land plots leased out in Beijing from 1992 to 2002 had been only 98.11 million square metres (Yu 2004). In the 2003 inspection of the national land market, the MLR noticed that 98 per cent of the land plots in Beijing were still leased by closed-door negotiations, of which 50 per cent turned out to be illegal (ibid.). In the designated green belt area, in particular, the area of construction land increased by 8.3 million square metres between 2000 (69.5 million square metres) and 2005 (77.8 million square metres) (BAUPD 2013). The role of green belts was hence looming large in legitimising and promoting the (local) state's land businesses.

More than this, the green belts were also deployed in consolidating the BMG and its affiliated companies' 'small treasury' as well as the business opportunities of related localities (villages or townships). On 6 August 2002, the General Headquarters of the BLGCGB declared that remaining construction land plots in the first green belt were all allocated to the 'Green Belt Infrastructure Development and Construction Company', an enterprise founded by the BMG in 2000 (BLGCGB 2002). While the advertised goal was to enable this company to provide adequate urban infrastructure for the green belt area – an aim mostly unfulfilled (BAUPD 2013) – its real concern was with the rising land interests: 243.22 hectares of construction land plots, all with a huge potential value, were occupied overnight by this BMG-owned company (BLGCGB 2002). On the other hand, to motivate the participation of villages and townships in producing the green space, the BMG allowed these localities to run the so-called 'green-based industries' on 3–5 per cent of its total green area. In total, 41 projects were approved in the green belts between 2000 and 2012, of which 23 were for games and sports (by and large, golf courses) (see Figure 3.1), five for leisure and vacation (resorts), six for ecological tourism,

FIGURE 3.1 A golf course in the planned second green belt

Source: Yimin Zhao

and seven for business apartments (BAUPD 2013). Under such a green mask, the number of golf courses in Beijing after 2004 increased from 20 to 70 (Du 2011), even though 2004 was also the year when the State Council (2004) halted the construction of new golf courses all over the country.

With the above discussions, it becomes clear that Beijing's green belts, an imported component of its modernist layout, were made into a powerful tool to contribute to the local state's land-based accumulation in the last two decades. This change was possible because the ideological connotation of the green belts was rewritten by the changing needs of the Party-state to make Beijing a 'modern and international' city. The ideological transformation was then practised to prepare for bidding for the 2008 Olympic Games, enabling both the institutional restructuring and political mobilisation at the city level. The designation of exceptions and the change of institutions, in turn, laid the political foundation for the booming land businesses of the state in the green belts. However, even under this political setting, it is still not clear why the large number of villagers who were subject to relocation when the land projects were unfolding, were compelled to accept the state actions. To understand this puzzle, it is necessary to see how and how far this set of institutions as a whole was turned into an effective ideology and preconditioned the success of the state-led and land-based accumulation.

From official urbanisms to the *only* way of life

As indicated at the outset of this chapter, politico-economic mechanisms underlie the formation and sustenance of official urbanism. For Lefebvre (2003), existing urban vocabularies that used to shape our conventional understanding of urbanism were compiled in the industrial age, while the urban problematic has already surpassed the industrial counterpart and become the predominant one. Hence, 'urbanism only serves to more cruelly illuminate the blind [between the industrial and the urban]', where 'the urban is veiled; it flees thought, which blinds itself, and becomes fixated only on a clarity that is in retreat from the actual' (ibid. 40 – 41). The gap between the institutional and ideological construct represented by official urbanism on the one hand and the politico-economic reality on the other is especially significant in China, where official urbanism shapes the urban mode of living into only one direction: urbanisation means modernisation (Li 2013). The Party-state aims to restructure the landscape of the peripheries in a completely urban way; for this aim, the lifestyle of local villagers needs to be remade in order to conform to the restructured landscape and urban imaginaries. This is the process in which the above set of institutions in the green belts are wrapped into an ideology and successfully instilled into the ethos of the population as the only desirable way of life in the urban age.

In the project agenda of Beijing's green belts, released on 20 March 2000, the BMG (2000a) claimed that speeding up the construction of the green belt area is significant for rectifying social disorder, facilitating the urbanisation process, and accelerating the sustainable development of the ecology, the economy and the society as a whole. In this pattern of social improvement, it continued, 'the rural mode of production and lifestyle should be urbanised, which is critical for the improvement of their life quality' (BMG 2000b). In this way, the green belts project was interpreted, ideologically, as an upgrade of the peasants' way of life. This marks the juncture where the two aspects of 'official urbanism' – institutions and ideology – were dialectically articulated. Concrete measures were immediately practised by the BMG to materialise such an ideology and hence to instill it as a belief of the peasants whose lives were to be fundamentally transformed thereafter. The measures could be summarised with two categories: relocation (安置, *anzhi*) and hukou upgrading (转居, *zhuanju*).

For the relocation of local villagers, various 'new village' projects were carried out with the endorsement of the BMG. Townships and village collectives were allowed to cooperate with private property developers and enjoy interest-free bank loans. In addition to building flats for the relocation of local villagers who were members of the village collectives, commodity housing units were also built in order to generate profits to finance the project costs and guarantee profits of developers. This was on condition that the total floor space of the commodity housing units was less than that of the relocation flats (BMG 2000b). According to a local official from BMCUP (interviewed on 1 August 2014), such procedures were defined as 'upgrading to the storied buildings' (上楼, *shanglou*), representing a physical transformation of a rural mode of living. On the other hand, the social welfare aspect was also attended to by the BMG, which was implemented through the process of '*hukou* upgrading'. The *hukou* system is indeed 'one of the most important mechanisms determining entitlement to public welfare, urban services and, more broadly, full citizenship' (Chan and Buckingham 2008: 587) in China. In the light of the Stalinist/Maoist ideology of industrial production, the system has long favoured urban citizens since its introduction in the late 1950s. Here, in the green belts project, the implicated local villagers were entitled to *urban hukou* status, thus becoming eligible to fully state-sponsored social welfare provision.

Drawing on the above two categories of upgrading (the physical and the social welfares), the BMG continued to declare that '[we should] fully respect the role of peasants as the *subject* of the construction of the green belt area' (BMG 2000b). This is the moment when the BMG tried to persuade the implicated population to accept the green belts project as the only channel towards a new and desirable way of life, and hence to establish a consensus on all related politico-economic institutions hidden behind the project. The persuasion succeeded in a straightforward way since it matched quite well with the desire of villagers who were also eager to change their living environment and the lifestyle in the process of urbanisation. For the area implicated by the green belts project was the same area where most of the migrant workers stayed when they arrived at the city. This demographic change in the local communities induced a specific socio-economic situation, in which

> The inflowing of so many migrant workers put great pressures on our infrastructure – environmental hygiene, electricity and water, and maintaining public order. It was common all over this area to find criminal activities

> spreading. Hence, we were feeling lucky that our old village got demolished quickly and our villagers relocated to storied buildings quite smoothly. Villages nearby, which are yet waiting for demolition, always tell us how envious they are.
>
> *(Interview with a village cadre)*

The above quote shows how far the official discourses and ideology were accepted by the villagers, who felt the negative impacts in their daily lives but did not take a single step forward to ask why. Instead of discerning the origin of such effects in the misconduct of the state, they were rather subordinating themselves to the official discourse. Since the built environment in reality was indeed dirty and messy, it was also their wish to transform it. But the only way they had worked out for doing this was the modernist spatial imagination of urban space that enables the state's land businesses to flourish in the green belt area. Both state agents and villagers advocated that 'urbanisation is the only way towards modernisation' (Li 2013) – and it marks the moment when the ideological consensus is established and the 'official urbanism' consolidated. This situation echoes Shin's (2014) recognition of the widely shared belief that the 'city makes life happier', a slogan of Shanghai EXPO in 2010, which articulates techniques of the state and desires of villagers, and which turns out to be the ideological foundation for the state's land business.

Such collusion between the state agents and the local villagers can be registered even more vividly via the responses of the latter in their everyday life. In a township called Sunhe located on the northeastern outskirts of Beijing's central districts, the green belts project became influential in the early 2000s when more trees were to be planted along the Airport Expressway in preparation for the 'Green Olympics'. One hundred and thirty-five households of villagers who lived along the expressway were relocated, with 490,000 Yuan or approximately £51,000 of compensation fees paid to each household.[3] Relocated villagers told their stories quite happily because it was before housing prices in Beijing had begun to rocket upwards, and they could buy a resettlement flat with around 220,000 Yuan, less than a half of the compensation fees they received (Interview with villagers in Sunhe). In other interviews, villagers revealed that conditions of life after relocation were better than the previous ones because of associated exceptional and privileged treatments (see Figure 3.2):

> The quality of my new flats is quite high. They are in a high-rise building, with more than ten stories, which even has lifts! This makes my life convenient because I am not agile at all after a surgery several years ago. Though the interior design is not the style I prefer most (with three double bedrooms), I am still quite happy with these new flats.
>
> *(Interview with a villager in Sunhe)*

> All facilities we can expect [in the urban life] were installed, such as the running water, the electricity, the natural gas, and the heating equipment. They

FIGURE 3.2 The street view of the resettlement community in Sunhe

Source: Yimin Zhao

> render the living here much more comfortable than our previous life in the old village. In addition, our *hukou* status is also changed into the urban category, which means we are now enjoying social welfares that are exclusively for the "urban citizens." However, the "soft environment" in this community is still unsatisfactory – but the main reason is that the peasants have not been dropped their old habits yet. For example, quite a few of them eliminated the lawn in the public space to plant vegetables. I think five to ten years are needed before them changing habits.
>
> *(Interview with villagers in Sunhe)*

In these acclamations of the new urban mode of living, the Party-state's official urbanism was successfully instilled into the common sense of the population and this in turn legitimated the former's political mobilisation for land businesses.

How much such seemingly enthusiastic responses from interviewed villagers were made with a clear understanding of their current and future circumstances is also yet to be verified. For instance, it is not clear if the level of villagers' participation in collective affairs in the coming years would decrease after land expropriation and relocation, following the patterns identified by Sally Sargeson (2016) in her study of five villages in Zhejiang. There is also the possibility that villagers' 'voluntary' move was associated with the implementation of successful preventive measures by the higher authorities using what Kevin O'Brien and Yanhua Deng (2017) refer to as 'psychological coercion' and 'relationship repression'. Also, the relocated villagers may incur greater expenditures while their actual income decreases, an experience of displaced farmers in Lynette Ong's (2014) study in Hefei. It is also not clear if the villagers knew how profitable the resulting land businesses by the local state turned out to be, and if the amount of distributed compensation, either in kind or in cash, was adequate. Other anecdotal evidence produced elsewhere suggests that there is a huge gap between the amount of land revenues and what is given out as compensation to villagers. For instance, a survey by an organisation called LANDESA in 2011 reports that 43.1 per cent of the surveyed rural households experienced land-taking, that 'affected farmers received some compensation in 77.5% of all cases, were promised but did not receive compensation in 9.8% of cases, and were neither promised, nor received compensation in 12.7% of cases', and that those compensated farmers received 18,739 Yuan as an average amount of compensation, which was only 2.4 per cent of average sales price earned by local authorities (LANDESA 2012). It is possible that the villagers interviewed above would have also been treated in a similar way but without their knowing.

What is evident though is that urbanism as an institution and an ideology has gradually obscured the boundaries between the legitimacy of the state, the logic of capital accumulation, and the mode of life of ordinary people during the urbanisation process. The use of official urbanism turns out to be one of the mechanisms through which the urban strategies of the state are implemented yet at the same time concealed, co-opting villagers into endorsing the state project.

Conclusion

In this chapter we have considered the use of green belts at the turn of the century by the Beijing Municipal Government to produce official urbanism that was utilised as a state project. We argue that ideological connotations of Beijing's green belts had shifted to juxtapose the concern for green space with the appeal to be modern-international. Drawing on the changing discourses, state institutions and policies were in turn rearranged to give priority to mobilising political resources for the successful bidding of the Olympic Games in the first instance; but it was also evident that the ideological and political manoeuvres were mostly directed to realising the economic ambitions of various state agents, who were facilitating land businesses in the name of conserving green belts. In this way, the green belts that originally articulated the socialist-utopian vision of the urban ecology were rendered a handy tool for the state-led and land-based accumulation in China's 'urban age'. While the focus of China's official urbanism has fundamentally shifted from industrial production (in the 1950s) to land businesses (at present), the nature and the role of this official urbanism have not changed: such urbanism works as a state project in which political mobilisation and economic ambitions are practised and consolidated through the urbanist discourses. As pointed out by Lefebvre (2003:

140), 'as an ideology, urbanism dissimulates its strategies. The critique of urbanism is characterised by the need for a critique of urbanist ideologies and urbanist practices (as partial, that is, reductive, practices and class strategies)'.

Scholars often highlight land-based accumulation as the predominant character of China's urban political economy and of its (neoliberal) urbanism (see for example, Lin *et al.* 2015; Lin and Zhang 2015). This is also evident in the story of Beijing's green belts, however this chapter has gone further to ascertain that China's urbanism is more than land-based accumulation. For urbanism as a concrete mechanism incorporates the whole process in which land businesses are initiated, endorsed, and facilitated in the ideological, social and political aspects. The economic interests (in maximising land revenues) mark the key concern of the Party-state and its official urbanism, but they are not a proper point of departure, nor the destination, of empirical explorations. The economic ambition can never be materialised in a political and social vacuum; instead, it has to go through the integration of ideological connotations, political mobilisation and territorial-scalar collusion/collision among various levels of governments. This politico-economic dynamic marks the concrete mechanism through which urbanism is shaped into, and deployed as, an institution and an ideology for the Party-state.

In addition, the nature of the official urbanism as both an institution and an ideology is indeed dialectical in the sense that, on the one hand, its ideological connotations constitute part of the set of state institutions, and, on the other, this set of institutions is then deployed as an ideology to reshape the belief of the people. The articulation between Beijing's green belts and the new urban political economy, for example, started from the moment when ideological connotations of these belts were rewritten in light of the Party-state's new needs. Here, political mobilisation took place in order to lay the foundation for generating and capturing land values in the green belts. At the same time, however, the success of the state manoeuvre was possible only when the affected villagers with entitlement to compensation were compelled to accept the official urbanism as the only promising way of their life and hence embraced the state conduct. This is how the official urbanism as an ideology works as constraints on protesters and as a facilitating mechanism to encourage consenting villagers in endorsing the state's ambitions. This recognition marks, for researchers in China urban studies, the potential added-value for the use of urbanism in understanding and analysing the great urban transformation of the country.

Notes

1 There are two green belts in Beijing. The first one was included in the city's master plan in 1958, and the second one was proposed in 2003. While the first belt embodies the import of the green belt as an idea and showcases its persistence in the master plan, the second belt is merely an unsuccessful mimicry with a certain politico-economic concern of the municipal authority (Zhang 2007). This will be discussed later in this chapter.

2 The air pollution issue was, of course, not the only factor that led Beijing's bid in 1993 to fail. A more critical factor was the issue of human rights, since it was just four years after

the Tiananmen Square crackdown in 1989. Human rights disputes induced geo-political pressures and both of these affected the whole lobbying campaign (Luo and Huang 2013; Riding 1993; Shin 2009a; Tyler 1993).

3 The calculation for the GBP from the Chinese Yuan here is based on the currency rate on 1 January 2016 (9.6123 CNY per 1 GBP). Source: Exchange Rates UK. URL: http://www.exchangerates.org.uk/GBP-CNY-exchange-rate-history.html. Last accessed: 23 September 2016.

4

NATURE, HOUSING AND EVERYDAY LIFE IN CHINESE CITIES

Junfang Xie

Introduction

There has been growing concern across the social sciences that urban studies has been overly dominated by theoretical and empirical perspectives developed with reference to a small number of 'archetypal' cities in Europe and North America, and that there is a need to undertake critical research which includes a more diverse range of cities around the world (Robinson 2002; Edensor and Jayne 2011).[1] Toward such a project, this chapter contributes to the burgeoning body of literature focused on critical engagement with urban change in China (see Chapter 1, this volume). However, while increasing numbers of studies have focused on a diverse range of topics and discussed a variety of case studies cities, Pow (2011: 47) argues that much of this research has fallen into the trap of assuming that 'political and economic effects peculiar to China have rendered the Chinese urbanization trajectory *more different* than similar from Anglo- American cities … [and accepted rather than critiqued how the] Chinese state in particular is seen to respond to and/or create conditions and institutions that render urban China's experience as unique and exceptional' (Pow 2011: 47). The focus of this chapter then is to contribute to 'unbounding' study of Chinese urbanism in order to develop a better theoretical and empirical understanding of the differences, similarities, connectivities, mobilities and relationalities within and between Chinese cities and those elsewhere in the world.

More specifically, I engage with the historic and contemporary development of Chinese cities with reference to changing political, economic, social, cultural and spatial issues bound up with nature, housing and everyday life. The relationship between growth of cities and the increasing need for housing and the associated social impact, either directly or indirectly has been a source of widespread academic and policy concern (Sandstrom *et al.* 2006). Moreover, it has been observed that urban sprawl has reduced green spaces in cities and resulted in a fragmentation of wildlife habitats (Swenson and Franklin 2000). Indeed, in many parts of China, rapid urbanisation has encroached into farmland and is enveloping villages and towns (Seto 2004; Schneider *et al.* 2005), with both 'planned' and 'wild' areas becoming 'islands of nature' amongst the urbanised landscape which itself has become more diverse compositionally and geometrically, and ecologically more fragmented (Zhang *et al.* 2004: 1 – 16).

However, while there has been a small amount of academic interest in positive and negative impacts on human life relating to nature and housing in China (Jim and Chen 2006; Xu *et al.* 2011), more attention has been paid to housing development more broadly (Wang and Murie 1999; Tomba 2005; Wu 2004; Zhang 2010). Such writing has focused on development of urban housing in relation to the emergence of state socialism, through the 1950s to 1970s, when the political and planning dominance of work-units (*danwei*) were responsible for providing public housing. Since 1978, an 'open door' policy has ensured a change from a 'national housing allocation system' to the building of 'commercial housing' dominated by the market economy, alongside a parallel provision of high-density, high-rise state housing (Wang and Murie 1999). More recently, spurred on by the real estate boom there has been an emergence of 'gated communities' throughout China which seeks to satisfy the needs of increasing numbers of middle-class residents who are looking to achieve 'traditional' notions of a 'good life' underpinned by both social and physical distinction which is in turn profoundly transforming urban landscapes (Zhang 2010). In particular, such transformations have been characterised by the emergence of an eclectic mix of 'Western' building styles becoming popular in cityscapes, as well as a re-emergence of an affinity with 'historic' Chinese design features which includes a re-articulation of connectivity with, and desire to 'live with nature' (Zhang 2010).

In this chapter, I draw together writing focused on theorising nature and everyday life in order to offer critical insights into urban political, economic, social, and cultural change in Chinese cities. I begin with a review of religious philosophies and cultural traditions relating to the development of Chinese gardens, and then discuss the impact of the introduction of 'Western' style public parks alongside the emergence of modern urban planning and design. The second half of the chapter traces how changing notions of nature and everyday life can be mapped onto housing development and associated urban social and spatial inequalities. In conclusion I highlight how theoretical insights at the intersection of thinking on nature and everyday life can inform socially progressive urban development, planning and policy and contribute to advancing critical perspectives on Chinese urbanism.

Theoretical context: nature and everyday life in the city

In order to understand the relationship between nature, housing and everyday life in Chinese cities this chapter draws on two bodies of critical writing. First, across the social sciences theorists have discussed the environmental basis of urban life and living, and how the 'urban natural environment' and 'nature' itself are subject to political, economic, social and cultural definitions, understandings and influences (Benton-Short and Short 2008). In these terms I address 'nature' as 'a contested concept' that deserves critical attention (Ginn and Demeritt 2009). I advance this perspective in order to engage with writing that has focused on the 'commodification of nature' in China, such as Li's (2012) study of Shanghai's Houtan Wetland Park that was developed as part of the Expo 2010, as a post-industrial site based on the cultivation of an urban 'wildscapes'. Also of particular relevance is Zhang's (2010: 91) work on the emergence of gated communities in the suburbs and central city which highlights historic motifs which 'make people who live [t]here feel they have returned to nature even though they live in the heart of the city'. Iossifova (2011) also describes how in close proximity, new urban gated communities have enveloped 'public space', greening 'fenced-off' areas of the city, in contrast to disadvantaged areas where 'nature' may only be glimpsed in window-boxes, vegetable plots and by the side of the road.

Such writing is complemented by studies of conflicts and tensions that circulate around 'nature' and urban living from elsewhere in the world (see for example Heynen *et al.* 2006; Whitehead 2005; Zerah 2007; Loram *et al* .2007; Shillington 2008; Woolley 2010; Beer *et al.* 2003; Pauleit *et al.* 2003; Bhatti and Church 2004). In particular, this chapter takes its cue from critical writing on urban natures in order to understand growing inequalities within contemporary Chinese cities specifically with regard to the quality of, and access to, green space amongst different residential communities (Zhang 2010).

Second, I draw on theoretical discussion around everyday life in order to understand the social and environmental inequalities bound up with access to 'nature' in Chinese cities. More specifically I seek to understand how 'expressions of ordinary injustice have a significant impact on the capabilities of disadvantaged urban communities to live out a full life' (Whitehead 2009). As Katz and Kirby (1991: 91) argue 'by comprehending nature, we reassert our power to reconstitute social nature, a power that is immanent in the practices of everyday life' (also see Bhatti and Church 2001). I also consider the 'fuzziness' and conflict relating to boundaries when it comes to conceptions and experiences of 'public' and 'private' with regard to green space in residential areas (Bromley 2004; see Chapter 2, this volume).

For example, Bromley (2004) argues that there is a continuum of 'public' and 'private' space, showing that 'gardening' is often an example of private encroachment onto public land where 'private' actions in the 'public' domain are often associated with 'informal' or even 'illegal' activity. However, Bromley (2004: 294) sees such boundary crossing as a 'moral' logic rather than 'taking possession of public space in a selfish way' (see Chapter 7, this volume). In these terms, it is important to consider the ways in which 'public' and 'private' are not exclusive or exhaustive categories but are socially and politically formulated and that 'popular meanings can be produced through dialogical encounters ... [where] respondents look to the material form of the site, and its location, in order to discern the intent of the space and thus shape a moral and aesthetic response to it' (Bromley 2004: 294). Such concerns will be drawn out with reference to the writing of De Certeau (1984) and Lefebvre (1971) who have highlighted the experiential dimensions of urban life and the significance of the city as a site of struggle and resistance between the powerful and less powerful. Indeed, Schmid (2012: 58) suggests that

> the point of departure for critical social theory should always be everyday life, the banal, the ordinary … changing everyday life: this is the real revolution … and any point has the potential to become central and be transformed into a place of encounter, difference and innovation

of a collective movement and asserting a 'right to the city'.

By drawing on theoretical perspectives relating to nature and everyday life, I also respond to Jayne and Leung's (2014: 1) criticism that, 'despite increasing academic interest focused on Chinese cities, there has been relatively little sustained theoretical and empirical engagement with critical approaches that have enlivened urban geography over the past 30 years'. As such, in order to offer critical insights into the relationship between nature, housing and everyday life in urban China this chapter begins by considering historical relationships between religious philosophies and cultural traditions relating to the development of Chinese gardens, and then discusses the impact of the introduction of 'Western' style public parks alongside the emergence of modern urban planning and design. The second half of the chapter traces how changing notions of nature and everyday life can be mapped onto housing development and associated urban social and spatial inequalities.

Gardens, philosophy and public parks: historic perspectives on nature and everyday life in Chinese cities

Classical gardens are highly valued as political, economic, social and cultural indicators of the trajectory of the Chinese nation from the earliest days of ancient dynasties to its contemporary growth as a global superpower. For example, traditional gardens represent pre-modern political, economic and social progression, technological and horticultural advances (M. Li 2009), and capture Chinese conceptualisation of nature as 'harmony of heaven and men' and 'letting things take their natural courses' (Hu 2013: 13). Such ideas remain key constituents of the importance of nature in everyday domestic life and notions of home and family today. More broadly the global importance of the Chinese classical gardens is widely recognised, as Jellicoe, president of the first International Federation of Landscape Architecture, wrote in the article *The Search for A Paradise Garden*, 'gardens of almost the whole world are based on the Chinese, the Western Asian, and the ancient Greeks' (M. Li 2009: 3).

The development of Chinese gardens has been described in term of five historic periods, with reference to political, economic and social and cultural change (An 1991; Hu 2013). First, during the 'germination' stage (1100 – 300 BC) of the Shang Dynasty, gardens were constructed by a privileged social class that emerged with capital accumulation generated by slavery. As well as being a site for the slave owners to pay homage to their ancestors, worship divinities and pray for immortality, gardens were a place for hunting and entertainment (Keswick *et al.* 2003). Garden symbolism represented nature as mysterious and beyond human control (An 1991: Huang 2008). Viewing platforms called '*Tai*' and '*Xie*' were built during this time, places where social elites could view the power and authority of gods such as *Xi* who lived in '*Yao-Chi*', which was the abode of immortals, a beautiful garden described as the ideal place for human life (Feng 1990).

Second, the 'growing stage' during the Qin and Han dynasties (300 – 200 BC) saw the development of 'imperial' gardens alongside the emergence of centralised feudal political and economic structures (Williams 1997). Gardens of this period were characterised by 'one pond with three celestial hills' designed according to ancient mythology, harmonising the spirit and human world, and in doing so inspiring loyalty to Emperors and being celebrated as spaces of family life and immortality (Feng 1990: 16). Gardens represented supremacy of imperial power and aesthetical taste, embodying 天人合 (*tian ren he yi*) 'harmony between heaven and man' (Feng 1990; Hu 2013). As historian Ji Xianlin highlights, connections between heaven (*tian*) and mankind (*ren*) and humanity (*he yi*) were core foundations of Chinese philosophies of Confucianism, Taoism and Buddhism (Watson 1958) where gardens 'although man-made, yet no less beautiful than the creations of nature herself' (J. Cheng 2012: 61). Gardens also drew on Fengshui to expresses human spirit, values and harmony and Buddhist pursuit of quiet (Liu 2011).

The third stage in the development of Chinese gardens was a 'transformation' period, associated with the rule of Wei, Jin and the Northern and Southern dynasties (220 – 589). This era can be defined in terms of war and political chaos as well as being a time of flourishing arts and culture, advancement in technology, and the spreading philosophies of Buddhism and Daoism. Emperors were keen to highlight their increasing economic and political power alongside expressions of their cultured lifestyle and patronage (Keswick *et al.* 2003; Wang 2006; Cheng 2012). Fourth is the 'bloom' stage during the Sui and Tang dynasties (581 – 907), acknowledged as a 'golden period' for Chinese gardens during the unification of warring dynasties and the emergence of China as a nation-state. During this time principles of garden design continued to be translated into other cultural forms and practices to a degree that 'all gardening is Chinese landscape painting … mountains-and-waters painting' (Feng 1990: 22). As the renowned intellectual An Huaiqi remarked in his writing, poetry, painting and gardening were the 'new Graces who dress and adorn nature' (Feng 1990: 22). Gardens were increasingly aligned with spirituality, symbolising a lifestyle removed from worldly temptations, and of mind-and-spirit freed from social worry. Such imagery inspired the popularity of private gardens constructed by an emerging literary, scholar, and bureaucrat class.

Finally, following 2000 years of development the 'mature' stage of Chinese historic gardens took place during the Song to Qing dynasties (960 – 1911). As explorer Marco Polo described, 'the palaces deserve to be ranked as the largest scaled construction in the whole universe … the enclosed garden could house extreme splendor and entertainment' (Hu 2013: 33). Chinese gardens thus reflected progressive social, cultural, moral and spiritual life expressed through planting, architecture and layout (Cheng 2012a, b; Hu 2013). In these terms, Chinese imperial, temple and private gardens have much to tell us about changing structures of political and economic power, social class, religion, aesthetics, kinship and family life. The ancient development of Chinese gardens highlights how 'urban natural environment' and 'nature' itself are subject to political, economic, social and cultural definitions, understandings and influences (Benton-Short and Short 2008).

The development of urban public parks

Prior to the dramatic removal of feudal regimes (1636 – 1911) China faced radical change in the latter half of the nineteenth century as a semi-colonial, semi-feudal society struggling to emerge from the First Opium War in 1840. Having shunned official contact with the outside world for two centuries the Qing government opened itself to (unequal) international trade through the Nanking (南京条约*Nan Jing Tiao Yue*) and Bogue treaties (Ma 2009). Through the influence of imperialist powers, European modernity proliferated in geographically enclosed urban settlement in selected Chinese cities, where wealthy foreign residents demanded a particular living environment and familiar urban facilities (Ma and Feng 2001; Zhu, 2012). For example, the development of urban parks in Shanghai mirrored their development in Europe to act as both the 'lungs of the city', facilitating healthy lifestyles to ensure productive workers, who for a short time could escape from everyday life in their 'slum' housing, as well as offering leisure spaces for middle-class Chinese and European urbanites (Zhou and Chen, 2009). The first 'public park' in Shanghai opened in 1851 (see Figure 4.1), and later was followed by a horseracing track with grandstands, formal 'gardens' with a lake for rowing boats and a bandstand (Chen 2001; Yang *et al.* 2003; Zhou and Chen 2009; Yang 2012). While the majority of local people were initially excluded from public parks, their popularity nonetheless soon grew. From 1868 – 1943 a total of 22 public parks were built in Shanghai, in parallel with a range of other important modern archetypal spaces and places such as public squares, theatres, cafés, dancehalls, department stores and so on built on the basis of European planning regimes and various European-style designed housing (B. Chen 2000). In contrast to imperial, temple and private gardens hybridised western/Chinese parks were key to proliferating ideas of 'public' urban green spaces (Zhu 2012; B. Chen 2000).

In the centuries before the emergence of the People's Republic in 1949, Chinese cites developed a rich and sophisticated tradition of imperial, religious and private gardens which relatively recently was augmented by the emergence of public parks. Understanding geographies of political and economic power, social and cultural change and religion and spirituality are key to unpacking the importance of gardens and parks and everyday engagement with nature in everyday life in China. However, with the introduction of public parks into China following the space, places and buildings that developed during the advancement of European

FIGURE 4.1 The first public park in Shanghai

Source: Shanghai History Museum

modernity it is important to think about the ways in which 'nature' became 'a contested concept' (Ginn and Demeritt 2009) that moved from being the sole privilege of wealthy Chinese citizens to being a key element in public urban design and planning regimes.

Urban planning, housing and green space

Departing from a mix-and-match of traditional Chinese and rationalised early-modern European planning, cities in China from 1949 onwards were dominated by late-modern Soviet traditions of planning and design, implemented in order to facilitate 'the great leap forward' achieved through rapid industrialisation. For example, the 'Five-Year Plan' (1953 – 1957) was underpinned by an integrated 'mode of urban living', based on provision of five or six hectares for each neighbourhood, and homes with an enclosed courtyard for the urban working-class which would provide a quiet living environment for residents, a design that echoed ideas of traditional Chinese gardens (see Figure 4.2). Moreover, following the first 'Five-Year Plan', Chairman Mao Zedong sought to 'turn the land green with parks and woods', including provision of woodland, public parks and vegetable and fruit farms in Chinese cities (J. Chen 2000: 10). However, in the push, and financial cost of pursuing accelerated industrialized urbanism, integrated provision of housing and green space urban living was not a governmental priority.

FIGURE 4.2 Cao Yang new residential district Shanghai, c. 1950

Source: Shanghai History Museum

Indeed, during the ten years of the 'Great Proletarian Cultural Revolution' (1966 – 1976) implemented to advance socialism and impose Maoist orthodoxy within the Party, chaotic dissolution of government organisation ensured that urban development stagnated. However, in seeking to solve serious problems of housing shortages major housing building projects were based on the construction of tube-shaped apartment housing (see Figure 4.3). With three to six floors without an elevator, organised along a long corridor of single rooms, no more than 20 meters square, residents had to share bathrooms and kitchens on each floor. Poor quality of construction and design of the building ensured that the great visions of Mao's 'green city' were lost in the economic instability and political inefficiency of the period (Katz and Kirby 1999).

By the late 1970s, the Chinese construction industry had nonetheless established higher quality standardisation and large numbers of public high-rise buildings were erected as the now dominant form of new residential development in contrast to older low-rise municipal city centres (see Figure 4.3). Around this time there was however a shift away from adherence to Soviet models of urban planning towards an interest in a North American model of vertical urbanism and suburban growth. A consequence of this policy transfer was a re-imagining of the link between nature and urban living. For example, in 1986, the first Chinese Garden Society Conference was held in Wenzhou and some forward-looking urban planners and designers began to pursue the concept of ecologically sensitive urban planning and development that draws on traditional concerns of closeness with nature, social well-being, spirituality and quality of life (Zhang 2010).

Of course, key to this beginning of a re-imagination of the importance of nature and everyday urban life were the sweeping economic reforms being implemented in China and the emergence of a market economy for housing related to the 'open door policy'. In 1982, Chairman Deng Xiaoping enabled the sale of public housing and adjusted rest structures in order to encourage residents to buy their houses.This partial liberalisation of the housing market enabled the diversification of housing development, and began to improve the quality and design of new development as stimulating 'green' planning regulations for both public and privately funded residential

FIGURE 4.3 Tube-shaped apartment housing, c. 1960

Source: Shanghai History Museum

FIGURE 4.3 (Continued)

development. However, such progress has taken place in the context of increasing house prices, rapid redevelopment of large areas of Chinese cities and lack of legislation and political will to control the commercial housing market. The role of the state in failing to regulate the newly formed privatised housing system and to provide high-quality housing for low-income families who have been increasingly economically and physically marginalised in Chinese cities has been the subject of significant concern. Thus, while the proliferation of gated communities, and medium- and high-rise residential developments over time (see Figures 4.5, 4.6 and 4.7) is continuing

FIGURE 4.4 Beijing's QianSanMen high-rise residential housing, c. 1976

Source: Shanghai History Museum

FIGURE 4.5 New urban residential district housing, c. 1980s

Source: Shanghai History Museum

FIGURE 4.6 Shijiazhuang private residential housing, c. 1992

Source: Shanghai History Museum

FIGURE 4.7 'Vanke' scene garden housing in Guangdong, c. 2000

Source: Shanghai History Museum

at a rapid pace in Chinese cities it is argued that it is the emerging new middle-class who are most able to live in urban residences with privatised green space in close proximity to their homes (Zhang 2010).

Nature, housing and everyday life in the contemporary Chinese city

Such social and infrastructural change must be understood in the context of unprecedented urban development in China. The Chinese National Bureau of Statistics showed that in 2008, there were 655 cities in China, 41 of which had a population of more than 2 million people and 264 of less than 200,000. However, the number of cities has increased by 151 per cent during the last two decades, while the total national urban areas have doubled. Demand for housing, especially commercial developments, has led to rapid increases in building programmes. While increasing in size and number, Chinese cities are nonetheless characterised by lack of appropriate legislation and planning control, and a mismatch of housing supply and affordability.

Within this context enthusiasm for the translation of traditional attitudes and values regarding the political, economic, social, cultural and spiritual value of 'living with nature' nonetheless remains a strong undercurrent. For example, the national Urban and Rural Planning Act of 1985 includes 'urban green structures' within city master plans. In echoes of the grand plans of Mao, current rapid and unfettered marketisation of urban housing is nonetheless clouding the implementation of such plans. For example, there are detailed and sophisticated plans for ensuring the provision of diverse green spaces at regional and urban levels and clear guidance on planning for both public and private green space with regard to diverse residential developments (see Tables 4.1 and 4.2). Despite government concerns to protect the relationship between housing provision and access to green space across socio-economic groups, encouraging the building of 'conservation orientated greening cities', it is the primacy of market forces that dominates urban development priorities (see Chapter 3, this volume).

It is with this backdrop of fast-paced urban change in China that we must develop theoretical and empirical insights into the details of how the ubiquitous spread of 'gated communities', populated by the emergent 'new middle-classes'; growth of high-density, high-rise state housing for low-income families; the development of public/private partnerships to fund the building of new urban parks (such as Xujiahui and Xintiandi parks in Shanghai) and the eclectic mix of 'western' building styles and social and cultural practices and lifestyles alongside the re-emergence of an affinity with 'historic' Chinese design features and a re-articulation of connectivity with, and desire to 'live with nature' is unfolding in Chinese cities (Zhang 2010). However, there is a paucity of research that looks in detail at everyday life for a diverse range of urban dwellers in the complex and diverse spaces of the contemporary cities throughout China. As such while the broad trends

TABLE 4.1 Typology of urban green space

Main typology	*Definition of green space*	*Main types of Green Space*	
G1 Parks Green space	to create green spaces for the recreation of the public urban population, including function of ecological, beautification, take precautions against natural calamities etc	Comprehensive parks	City's parks, community park, etc.
		Theme parks	zoo, botanica1 garden, historical garden and park, etc.
		Linear park	River and Canal Banks ,Transport Corridors (road, rail, cycleways and walking routes)
		Roadside green space	Street Plaza green space, a small street green land
G2 Productive plantation area	to provide green space for urban nursery-grown planting, flowers and grasses, and woodland for greening etc.	Nursery stock, flowers, seed nursery, tree nursery and grass garden.	
G3 Green area for environmental protection	green space used for urban environment protection, sanitation, safety and calamity prevention, including:	health segregation, road protective greenbelt, urban high-pressure corridor green belt, medium-thick shelterbelt and city group segregation, etc	
G4 Attached green space	the various types of land affiliated green space without the urban construction land	residential land, public facilities lands, industrial land, storage land, land for traffic, square of road, municipal facilities lands and land for special use, etc	
G5 Other Green Space	green space for quality of the city's, health and well-being of residents life, ecological environment and biological diversity protection, including	landscape and famous scenery, water reserves, country parks, forest parks and nature reserves, woodland, natural preservation areas, wetland, wild zoo and botanical garden, landfill recovery green space, etc.	

Source: Classification Standards of Urban Green Space' CJJ/T85–2002)

identified in this chapter offer important insights into Chinese cities there is a need for sustained empirical research to explore the everyday realities for the place of nature in the everyday lives of different social groups, in different urban spaces and places, in different cities throughout China.

TABLE 4.2 Classification of residential area and public and private green space

Residential Buildings	*Road and Pathway*	*Green space*	
High residential building(≥10floors)	Residential district road (up to 20m)	Open (public) green space	Residential district park (Outdoor play areas, Children's playground, fitness venues)
Semi-high-rise building (7–9floors)	Residential quarter road (6–9m)		Residential quarter recreation green space
			Housing cluster green space
Multi-storey building(4–6floors)	Housing cluster road(3–5m)	Enclosed (private) green space	Green space between housing
Low-rise dwelling Building(1–3floors)	The pathway between housing (up to 2.5m)		Green space of building foundation (For the building foundation and protection of green space)
			Linear green space of pathway

Source: Code for Design of Green space in Resident Areas, DB11T214/2003, Beijing Municipal Bureau of Parks

Conclusions

By drawing on diverse theoretical debates I have highlighted how the relationship between political, economic, social, cultural and spiritual discursive constructions of nature must be understood as an important feature of historic and contemporary Chinese cities. Political and economic priorities have been shown to be at the heart of both successful and unfulfilled aims to 'green' Chinese cities. Moreover, thinking about relationships between nature and everyday life highlights similarities, differences, connections, mobilties and relationalities between Chinese cities and European, Soviet and North American urbanism, as well as ancient religious and spiritual philosophies and practices with a long genealogy of Chinese history that is beginning to be re-imagined in the contemporary city. Given rapid change, and the diversity of Chinese urbanisms there is nonetheless important work to be done to better understand how broad urban processes are playing out in particular ways in a variety of spaces and places, amongst different social groups and in different cities. It is vital that such critical perspectives are pursued further so that theoretical insights and empirical evidence can contribute towards urban policy and planning agendas that seek to ensure more equitable access to, and increased 'quality' of green space in residential areas, as well as highlighting the importance of Chinese urbanism to cutting edge international urban research agendas.

Note

1 This chapter has been previously published as Xie, J. (2015) 'Landscape design, housing and everyday use of green space in urban China, *Geography Compass,* 9 (1), 42 – 55.

PART II

Identity, lifestyle and forms of sociability

5

ENCOUNTERING STRANGERS

Prostitution and urban life in Dongguan, China

Xiaomei Cai and Xiaobo Su

Introduction

Prostitution has become an important feature of theorising Chinese urbanism in recent years alongside increasing interest in bodies, gender and encounters. The development of commercial sex as a social-economic formation is an open secret in China although the sex industry is illegal. Transient encounter via commercial sex produces 'uncanny' intimacy that is romantic and repressive, and plays out in complex and heterogeneous ways in private and public space (see Chapters 2, 7 and 14, this volume). Encounters between female prostitutes and male clients effect meaning and experience of urban life, by involving commercial transactions, female chastity, and immoral bodies. Sex and love in the private sphere is of course the antithesis to commercial sex. This antagonism is aptly captured by Simmel (1950: 122) who argued that 'the nadir of human dignity is reached when what is most intimate and personal for a woman ... is offered for such thoroughly impersonal, externally objective remuneration'. For female prostitutes in Chinese cities, commercial sex affects diverse moments of everyday life and these encounters have been widely ignored in academic studies.

This chapter thus responds to this research lacuna and advances theorisation of female sex workers' everyday life in Dongguan – China's so-called sin city. We use the term 'encountering strangers' to denote conditions where sex workers have transient, but intimate, encounters with their clients, their friends, and the city itself. Despite the centrality of 'strangerhood' to practices of everyday life, or what Ossewaarde (2007) calls a 'society of strangers', the making and unmaking of strangers has remained largely absent from the literature on everyday life. Here we follow Ossewaarde (2007: 368) to define a stranger as a 'nameless human being without social characteristics'. In its self-definition, 'strangerness' is an ethos of 'modern' citizenship in a highly mobile world, a way of thinking about the relationship between individuals who are connected via transient encounters, and hence, as a way to interrogate relational identities. With this definition in mind, we look at how commercial sex turn these workers into 'strange' subjects and how workers engage with and experience 'strangerness' in their everyday life. By doing so, our analysis responds to Pratt's (2005: 1072) call for recognition of 'a body beaten beyond recognition' in order give voice to the disenfranchised in Chinese cities.

Encountering strangers in the city

Strangers have always constituted urban life. In Simmel's (1950: 402 – 403) theorisation, the stranger is an essential element of urban society and his or her position as a full-fledged member involves 'both being outside it and confronting it'. He or she intends not to stay long enough to become acquainted, or to seek to be included in local citizens' lives. Simmel (1950: 403) further argues that the stranger owns both nearness and remoteness, giving the stranger 'the character of objectivity'. In contrast, Schutz (1944: 506) argues that a stranger constructs 'a social world of pseudo-anonymity, pseudo-intimacy, and pseudo-typicality'. This 'disintegration' generates an in-between situation – remoteness and intimacy, hesitation and uncertainty – through which social relations are complicated.

Drawing on Schutz's conception, Ossewaarde (2007: 370) goes further to illustrate the key feature of a stranger where

> he or she is constantly confronted with the discrepancy between the local situation and the homely world of his or her past. Through this awareness of strangerhood, the stranger is not only a stranger for locals, but also becomes a stranger to the self.

This illustration deserves further elaboration partly because it shows a juxtaposition between 'homely worlds' and public spaces of the city (see Chapters 2, 7 and 9, this volume). To 'locals', the stranger that Schutz conceptualises is not a subject of esteem and objectivity, but a marginal person, an external intruder, who stays to make trouble and can be considered a challenge to social cohesion. Similarly, Bauman (1997: 28) highlights depictions of the stranger as hateful and feared: 'the acuity of strangerhood, and the intensity of its resentment, grow up with the relative powerlessness and diminish with the growth of relative freedom'. This point is advanced by Valentine (2008) who argues that scholars have often neglected the knotty issue of inequalities when celebrating the potential of everyday encounters in cities.

The image of a stranger as a trouble-maker and as an external intruder can be explicitly applied to sex workers in China. Indeed, what characterises female sex workers as strangers in China is contested notions of (im)morality. For example, female chastity has been a prominent narrative in Chinese society for centuries and reached its climax in the Qing dynasty (1644 – 1911) where chastity could bring families 'status, social leverage, and sometimes state honors' (Theiss 2004: 13). While Maoist revolution initiated female emancipation against the customs of feudal patriarchy, sexuality and female bodies continued to be 'the bearer of sexual and family morals' (Steinmüller and Tan 2015: 15). More recently, in post-Mao China, attitudes towards virginity and female chastity have been rapidly changing, as many young people both pursue romantic love and accept premarital sex. With the backdrop of changing attitudes female chastity has not, however, lost its symbolic value and the ideal of 'a virgin bride' still remains a key value for many in China (Steinmüller and Tan 2015). In contrast to highly prized 'virgin brides' sex workers of course are the antithesis of female chastity. Viewed as 'unruly whores' or 'fallen women' sex workers lives are infused with shame and discrimination in everyday life (Zheng 2009). Meanwhile, the Chinese state has launched numerous campaigns against prostitution under the guise of maintaining social morality. Linking prostitution and sexual servitude with morality alludes to a fundamentally Confucianism-oriented sexist position (Zheng 2009).

Bauman (1997) reminds us that all societies generate strangers, but different societies produce diverse kind of strangers. As far as sex workers in China are concerned, experiences of migration which already renders them strangers in cities are exacerbated by moral and legal conventions. Moreover, these strangers corrupt neat divisions between intimacy and unfamiliarity, between legality and illegality, and between innocence and impurity. For many in China, prostitution generates uncertainty and breeds 'the discomfort of feeling lost', while the sex industry defies the most ardent efforts made by the state to prevent or outlaw commercial sex. In other words, sex workers, as contemporary strangers in today's China, are a 'waste of the State's ordering zeal' (Bauman 1997: 18). Indeed, there have been numerous crackdowns on prostitution in Dongguan and other cities so that the Chinese state can pursue the goal of building a harmonious society.

Connections between space, place and sex work have been explored elsewhere in academic literature (Laing and Cook 2014). In particular, scholars have examined the socio-spatial exclusion and marginalisation of sex workers and the disciplinary power imposed on them (Hubbard and Prior 2013; Davidson 1998). According to Hubbard *et al.* (2008: 137), urban authorities in Europe endeavor to repress spaces of street prostitution and intervene in sex work markets to address gendered injustice, while actually reinforcing 'geographies of exception and abandonment'. However, it has also been asserted that there is a need to avoid regarding prostitution as 'one monolithic, unchanging institution' (Brents *et al.* 2010: 10), which overlooks sex workers' changing life experiences (Kempadoo 2001). To understand this diversity, we use the term 'encountering strangers' to analyse sex workers' transient, but intimate, encounters with others in their everyday life. Here encounter becomes a key feature to conceptualise urban life in Chinese cities, or as Valentine (2008) asserts, cities constitute a field of encounters with strangers. In these terms, new forms of coexistence and sociability and new sources of prejudice and tension emerge to shape the meaning of urban spaces (Wilson 2016). Indeed, in Chinese cities, sex workers constitute a key source of prejudice during their encounters with clients, residents, and local authorities. In Bauman's (1997: 30) words: 'the question is no longer how to get rid of the strangers and the stranger once and for all, or declare human variety but a momentary inconvenience, but how to live with alterity – daily and permanently'.

Being a stranger in Dongguan

Economic reform following 1978 activated a rapid pace of industrialisation in China's coastal cities, and thus induced tens of millions of rural-to-urban migrants to seek new job opportunities. The Pearl River Delta in general, and Dongguan in particular, have captured the attention of rural migrants, including young females who enter into the sex industry, giving rise to Dongguan's high profile as a sin city. In 1997, the Asian financial crisis inflicted heavy losses on Dongguan's export-driven economy. Numerous factories went bankrupt, and female workers lost their jobs. To make a living, some jobless women entered the sex industry. As many factories started to work with the International Organization for Standardization (ISO) to import various international standards and enhance their products' competitiveness in the global market, the sex industry, accordingly, redefined its own services by synthesising Thai massage, Japanese adult movies, and Chinese traditional acupuncture with the so-called Dongguan ISO aiming to maximise clients' sexual satisfaction. The 2007 global financial crisis later shattered Dongguan's export-driven economies again and more factory workers were drawn into the lucrative sex industry.

Our interviews with hotel managers reveal that large hotels' sauna departments often attracted young female sex workers. These women receive professional training, from two weeks to one month, with a focus on expertise in 36 service programmes with the goal of becoming Dongguan ISO-qualified technicians. These high-end workers offer a wide range of services to suit clients' preferences with the cost ranging between 600 and 1000 yuan for two hours. Higher-paid technicians could be paid 1500 to 2000 yuan for two hours. Each luxury sauna organised over 100 technicians for business per night. However, while Dongguan is reputed to have around 100,000 sex workers, only a small fraction work in luxury saunas (*Beijing Youth Daily* 19 February 2014). High-end sex workers in saunas earned much more than other types of prostitution such as street workers, service girls in karaoke or massage parlours, bar hostesses, and prostitutes working in brothels.

With this diverse background in mind, we wanted to know more about the ways in which sex workers negotiate and experience 'strangerhood' in their everyday life in Dongguan. To address this question, we adopted a women-centred and lived-experience approach to understanding our respondents' everyday lives (Kong 2006). From January to September 2013, we interviewed high-end sex workers who worked for a luxury sauna affiliated with a five-star hotel in Dongguan. The interviews with sex workers in Dongguan were focused on their everyday encounters beyond their workplace. Interviews were undertaken in Mandarin and lasted between 45 and 90 minutes. All respondents were given a pseudonym.

Pain and intimacy in the sauna

High-end sex workers' daily lives are highly routinized. Normally, they go to work around 6 p.m. in the sauna until 3 a.m. or 4 a.m., if they do not secure a 'one night package' client. Our respondents then often had a snack before going to sleep for seven to eight hours in their apartment either rented by themselves or their sauna manager. After a lunch around 1 p.m., our respondents usually had free time in the afternoon for shopping or spending time with friends. There was not time off at weekends or structured holidays unless permission from 'sauna mamas' was gained. Despite some reluctance to discuss social and economic interactions at the sauna, several different respondents did discuss their encounters with clients indicating their vulnerability in the process of transactions:

> At the beginning, I liked this sort of life since I felt free and happy. Now I realize I don't like it very much. Why? I met some abnormal clients. They tug my hair and beat me. Last month an old man was unable to ejaculate. I said to him: 'It is none of my business; it is yours'. Then he hit me and slashed my face. I was so scared.
>
> Protection? It is an ideal word. I don't even know how to protect myself. What I know is that the sexual encounter between myself and clients is a transaction. If some of my clients don't want to use a condom, I have no opinion. After all, I want to serve as many clients as possible. Regarding venereal diseases, how can I avoid them? Some days, I serve six to seven clients.

These quotes shed light on saunas as a working space where sex workers' encounters with unfamiliar clients are full of risk and unhappiness. As well as moments of violence, respondents discussed physical and psychological harm related to frequent unprotected sex with clients.

Some scholars have argued that commercial sex takes place in a paradoxical space that is 'imagined in order to articulate a troubled relation to the hegemonic discourses of masculinism' (Rose 1993: 159). In Dongguan, however, the sauna does not entail such a paradox, as it reifies an imagination of erotic femininity and long-established masculine cultures in China. In the sauna, vagaries around money and masculinity dissolve, simply because male clients are absolutely prioritised. Similar to night bars in Singapore (Tan 2014), luxury saunas in Dongguan encourage sexual discovery, fulfil sexual desire, and more importantly, pursue a logic of capital accumulation. However, in contrast to Tan's (2014: 42) observation that female clubbers use dancing to cultivate 'feelings of (sexual) empowerment, affect/agency, choice and intention', sauna sex workers' bodies and erotic femininity are regulated by the cultural logic of masculine dominance and the commercial objectives of profit making. Our respondents nonetheless highlighted a challenging juxtaposition – strangerhood and intimacy, nakedness and privacy, force and willingness – giving rise to an interstitial intimacy (see Chapter 7, this volume).

While our respondents were unhappy to work at saunas, they nonetheless preferred them to other workplaces such as massage parlours or karaoke halls. Local authorities in Dongguan policed and prosecuted commercial sex work, with police selectively raiding locations and venues in response to political or media attention, or in order to gain bribes from arrested sex workers and clients. Nevertheless, luxury hotel owners and sauna managers in Dongguan usually possessed the necessary connections to protect their saunas from being harassed by the police and illegal gangs. The saunas thus provide sex workers with some sense of security to serve clients without worrying about law enforcement. Furthermore, the saunas were a location where sex workers were more able to remain anonymous, as one respondent suggested: 'I didn't want to do business in KTV as I am afraid of bumping into acquaintances. I went to the sauna, a relatively secret place. The transaction has more privacy. Of course, the money is much better than in other places'. An understanding of pain and trauma notwithstanding, for our respondents the sauna was paradoxically a space of security and privacy, and an ironic form of 'protected intimacy' (Bachelard 1964).

Friendship, alienation, and a home away from home

In Dongguan, several sex workers normally rent an apartment and live together. By dwelling with co-workers, our respondents sought to evade the ubiquitous presence of demonisation and stigmatisation in the city. They allude to 'a new home' to frame their subjectivity and to enable more 'comfortable' social relations in Dongguan. What then constitutes a new home? Various interviewees identified friendship as a key element.

> I have a very good relation with three other sisters and we live [in one apartment] together. I don't talk to others very much. So, I pretty much remain silent.
>
> I have few friends here in Dongguan. No one knows me and I don't want to know others.
>
> I know two sisters. We often gather together to talk and encourage each other. They give me much positive energy and empower me to live on. They also buy presents for my son. They sincerely communicate with me and listen to me. I feel so grateful for knowing them.

As these narratives show, attachments to friendship reduced estrangement in their everyday life and fostered a sense of home in the city through comfort and familiarity, in order to withstand alienation and strangerhood.

In addition to socialising with friends, popular activities 'at home' included playing cards and mahjong, make-overs, and discussing fashion and branded commodities (see Chapter 8, this volume). Other activities included surfing the internet and watching romantic dramas. Nevertheless, depictions of love and romance often led to uncomfortable and challenging reflections; 'I love the internet and watching movies. Nevertheless, the love stories in dramas are so impressive that when I watch them, I feel I am so miserable.' Moreover, as another respondent suggested, being a sex worker 'closed the door' to a real love relation with clients or other males in Dongguan, as she and other workers have been labelled as 'dirty', 'polluted', or 'filthy'. Indeed, several respondents shared stories of transient romances with their clients, and that while the romances had ended they all emphasise that they felt respected by their lovers. However, in spending their leisure time online and in watching television dramas, our respondents highlighted social isolation because of being sex workers. Others discussed how gambling and alcohol were often used by sex workers to 'anaesthetise themselves' from physical and mental harm.

Consumption, economic and social power

Although enduring social stigma and isolation, sex workers nonetheless expressed the value of economic power to display 'femininity', particularly via shopping in Dongguan, and by visible signalling of luxury brands (see Chapter 8, this volume). Generally, respondents went out together to do shopping for luxury goods in high-end stores or to eat in top-quality restaurants. Indeed, several of the respondents refused to hide sex-worker identities, and would wear 'sexy outfits' and bright red lipstick when shopping; as one respondent complained, 'I didn't do anything wrong, right? I didn't steal nor rob others. Neither did I rely on my family. I earn money by myself. What is wrong? It is my job and here is my money.' Moreover, other respondents emphasised their contribution the local economy; 'Dongguan cannot develop so well if we didn't work here'; with other comments including 'my spending helps so many people, such as taxi drivers, property owners, and even sauna security guards. Without us, how can they make money?' Such comments reveal a sense of pride in contributing to developing the urban economy, and a balancing of social stigma and economic value by our respondents as they discussed their involvement in the sex industry and sought to contest stereotyped categorisations imposed on their subjectivity. This exercise of money is reminiscent of Cangiani's (2011: 182) argument that individuals can often resort to arguments relating to rational choices to maximise economic return when in market societies 'money becomes the means for day-to-day living and the medium of social relationships'.

Consumption was a key way in which sex workers embodied their subjectivities as 'urban and modern' as one respondent revealed:

> I purchase handbags, clothing, cosmetics, smartphone, computer and so on. My classmate often asks me to go shopping. She spends way more than I do. As a mistress, she has plenty of money. Being a mistress is better, isn't it? Nevertheless, I enjoy my current condition because I earn money by my labor.
>
> *(Interviewer:* Could you consider a different job?)
>
> Why should I do a different job? Shall I go to a factory? No way! How can I live [with a meagre salary]? Actually, I like my current job, though sometimes I feel exhausted. But, there is no easy job in the world. Now I have money to buy whatever I want. I feel quite good. I will work hard. The more money I earn, the more branded stuff I will buy.

Sex workers purchase of branded commodities allowed participation in consumer culture Purchase of expensive cosmetics, high-heeled shoes, fashionable clothes, smartphones, and gym membership, allowed sex workers to emulate professional women (see Chapter 8, this volume). Through consumption, our respondents transformed themselves from 'rural country girls into sexy, modern and urban women' (Ding and Ho 2013: 43). Such thinking resonates with Ding's (2012: 172) view that consumption enables sex workers in Dongguan to improve 'self esteem, and a form of self expression that may bring positive and pleasant feelings'. Respondents' ability to engage in consumer culture did not however negate the hatred and fear bound up in local residents' response to their presence. Some residents expressed overt hatred against sex workers 'as fallen women' and reprimanded them for corrupting their husband or partner because of their 'erotic femininity and sex skills'.

By focusing on the everyday life of sex workers, we have found that strangerhood and intimacy are paradoxically enmeshed in spaces of the sauna, 'home', the street, shopping mall and so on. Our research reveals complex webs of power that integrate respondents into the sex industry for economic profit, but entrench social and spatial barriers which are exclusionary and marginalising. Specifically, the sauna enables economic freedoms, and separates both sex workers and clients from a normative home life, representing a condition 'without duration' (Bhabha 1994: 204). Sex workers' bodies become public, for sale – and despite economic power enabling consumption as 'the source of pleasurable experience and aesthetic satisfaction' (Bauman 1997: 34) this intermeshing of intimacy and strangerhood, separateness and connection ensured 'an embodiment of difference' (Shields 1992: 189). In the workplace and 'at home' however, sex workers found friendship and mutual support – a sense of comfort – where 'the sensory experience of feeling at home … defies the insecurity and uncertainty that can come when confronted with difference' (Butcher 2010: 34).

Being a stranger forever?

Some sex workers chose to quit the profession and return to their hometown or a nearby town to enable a different life. To end this chapter, we briefly reflect on sex workers' experiences of such home-making. It is perhaps not surprising that there were mixed stories of the successfulness of social and spatial transitions. For some, their 'cover stories' of working in Dongguan, as well-paid nannies to rich families, had remained intact for many years, and return to hometowns or provinces led to a move away from prostitution, happy family lives and successful employment or business ventures. Others had been less successful in maintaining such 'cover stories' when working in Dongguan, had become estranged from family and friends and were subject to suspicion and malicious gossip when trying to forge a new life away from prostitution. Such stories highlight the challenges of living in the shadows of commercial sex and female chastity, and the sophisticated strategies that our respondents developed to 'in-between identities' or 'double lives' (Bhabha 1994: 26) or in response to encounters that don't take place in 'a space free from history, material conditions, and power' (Valentine 2008: 333). Personal, social and cultural geographies of Dongguan's sex industry continue to impact on everyday life often long after our respondents had stopped working as prostitutes. However, as research on sex workers in Dalian, a city in northern China, Zheng (2009: 157) highlights, a return to 'home', cities or provinces can be 'a place of refuge both physically and emotionally'.

Conclusion

In this chapter, we have presented research that investigates the everyday lives of sex workers in Dongguan – China's sex city. Departing from tendencies to romanticise urban encounters and to uncritically celebrate social difference in cities, we have offered theoretical and empirical insights into inequalities and power relations bound up with spatial and temporal intersections of gender, masculinity, femininity, sex and sexuality, consumption, economic exploitation, employment and so on. By developing understanding of how prostitutes in Dongguan are defined in relation to 'strangerhood' we have thus highlighted 'a collective experience of powerlessness, manifested in feelings of personal meaninglessness, loneliness, mistrust, insecurity and anxiety' (Ossewaarde 2007: 385). However, in returning to Valentine's (2008) challenge to critically interrogate the socio-spatial inequalities that people 'at the margins' experience in their everyday lives, the respondents in our study nonetheless clearly highlight how they develop their own 'critical urbanisms', which to differing degrees and in diverse and sophisticated ways, seek to respond to and overcome the inequalities, challenges and insecurities of their current and future lives.

6

GREENING THE CHINESE CITY

Young people, environmental activism and ChinaNet

Alison Hulme

Introduction

Academic engagement with Chinese cities has often been concerned with the rapid processes of industrialisation and urbanisation at play, including mass migration and the wealth divide between those in urban and rural areas (see for example, Weiping Wu 2012). Environmental consequences of these processes have also been analysed at length and there is a relatively rich and growing body of work on the role of non-governmental organisations (NGOs) in the burgeoning Chinese environmental movement (see for example Yiyi Lu 2005). Contemporaneously, significant attention has been paid to the internet in China (or rather, ChinaNet),[1] including its potential for the creation of a 'public sphere' (see for example, Rauchfleisch and Schafer 2015; Chapter 2, this volume). What has not yet received much attention is the role of the internet *as part of* the burgeoning environmental movement, and for example, its role in encouraging and facilitating environmental concerns, and specific campaigns and protests. In addition, the place of these environmental movements in the lives of young people in Chinese cities and the implications this has for discussions around Chinese urbanism has also received little theoretical attention.

This chapter focuses on the role of ChinaNet in the growing number of environmental protests in Chinese cities, and the way in which green issues are increasingly becoming part of everyday urban life and contemporary youth culture. In particular, I explore the ways in which protests in a specific city have provoked copycat activities elsewhere (such as the PX protests), creating environmental movements and uniting disparate cities and young people in a sense of belonging to what could be called a new progressive urban sensibility across China. This sensibility is outward looking and globally aware, as it is to some extent informed by mechanisms and policies of organisations and events such as the United Nations, the World

Trade Organisation, the Rio summit, etc. The suggestion in this chapter is that this sense of urban identity is understood as 'modern', not due to being 'green' (as there is plenty of evidence of environmental concern embedded in traditional Chinese culture – see Chapter 4, this volume) but through the act of campaigning in new ways, i.e. not the traditional petitioning of the Mao (or even the Deng) generation.

Chinese urbanism

By way of setting the scene for what follows in this chapter, it is important to note that Chinese urbanism tends to be discussed in relation to the vast array of economic and social changes that have taken place in China since the beginning of the leadership of Deng Xiao-ping in 1979. Not only did such changes see an increase in standards of living, mass migration to the urban, coastal areas, and rapid processes of industrialisation and urbanisation, but it they also heralded cultural change in cities. Deng placed emphasis on the creation of a *xiaokang*, or 'functionally well-off class',[2] for whom 'getting rich' was no longer shameful as it would have been under Mao's rule. As I argue in this chapter, the *xiaokang* is 'an urban class' due to access to wealth accumulated in cities, which has become far more than simply an 'economic class'; this new 'functionally well-off class' increasingly displays certain common cultural elements such as a tendency to place high value on individual and broadly liberal beliefs.

The emergence of the xiaokang is inextricably linked to the dismantling of the *danwei* or 'work unit' that existed under Mao. A danwei was like a very small walled, close-knit town, and was the principal form of social organisation (and of implementing party policy), providing employment, housing, food, clinics, childcare and schools, as well as social respectability or 'face' (*mianzi*). Allegiances were to one's community within the danwei; therefore the dismantling of the danwei has caused some commentators to argue that there has been a shift from the *danwei ren* (danwei person) to the *shehui ren* (social person), which corresponds to the rise of the market and the privileging of individual interest. In line with this, Yong Gui *et al.* (2009) assert that this separation of work space and living space, and the fact that young urban people now live in neighbourhoods where many are home-owners and there is a variety of income levels means that urban residents are less likely to feel an affiliation to a community (see Chapter 7, this volume). Tang Xiaobing also draws upon the difference between the danwei and 'the social person', suggesting two distinct responses to everyday life in China: as something to be overcome by a heroic commitment to communal living (the danwei), or, as someone who attempts to transcend the everyday and ameliorate anxiety through the consumption of lifestyles and commodities (the social person) (Tang 2002: 129–130). What is crucial here is that the shehui ren is very clearly a typical member of the xiaokang, and despite an argued lack of community allegiance, they may well seek other, wider affiliations that can be seen as part of a sense of belonging to an urban class or at least an urban *type*. Despite being a somewhat crass delineation of subjectivities in current-day China, this framework for thinking about urban culture does nonetheless serve to understand at a top level the nature of urban culture and why it might specifically foster grassroots movements that the mainly young participants feel connect them to other urban residents across China, and indeed across the globe.

'The environment' and Chinese young people

Ecological awareness has a long history in Chinese culture, being embedded in Daoist and Confucian thinking, for example, with the concept of the cyclical economy (a notion that finds form in many contemporary Chinese environmental policies, such as recycling; also see Chapter 4, this volume). However, an environmental movement, understood in terms of a unified and public set of concerns that guide policy and action, came to China somewhat later than it did other countries. As Arthur Mol (2006: 30) asserts, while in most industrialised countries the institutionalisation of the environment in national politics and policies had started in the late 1960s or early 1970s, in most developing countries this process began only in the late 1980s and early 1990s, and in China it did not arrive until well into the 1990s. This was partly as a result of China itself being part of the second generation of 'Asian tiger' economies (Malaysia, Thailand, and most recently Vietnam), as opposed to the first generation (Taiwan, South Korea, Singapore). It was also, as Mol argues, due to 'ecological modernization', at least until the mid-1990s, typically being seen as a Western theory whose major assumptions were ill-fitting for non-Western nation-states (Mol 2006: 30; also see also Frijns *et al.* 2000; Sonnenfeld 2000).

There is also an argument that China's industrial development under Mao was particularly damaging to the environment compared to the development of other nations, and that China was inclined not to *recognise* this damage, due to its determination to rid itself of the Confucian ideal of 'harmony between heaven and humans' in favour of the insistence on 'the people's' will to conquer nature. This is vociferously argued by Judith Shapiro in *Mao's War Against Nature* (2001), although it is debatable whether the Maoist insistence on human agency was any greater because of a lack of sympathy with environmental causes than a simple materialistic desire to develop and become wealthy was in non-communist developing countries. Indeed, this counter argument is borne out by the idea that it is precisely the fact that China developed so quickly, entering the global market with such force, and the 'wild capitalism' that ensued, that caused the scant regard for sustainable development in comparison to other nations.

Nonetheless environmental concerns did begin to emerge in China, from the mid-1990s, largely evidenced, at that time, by the creation of home-grown NGOs such as *Friends of Nature* and *Global Village*, and the allowing-in of foreign NGOs – the most famous of these being Jane Goodall's *Roots and Shoots*[3] Such organisations however, were, and are, limited in number and tend to position themselves as expert or awareness-raising organisations, as opposed to adversarial or confrontational pressure groups. As Mol (2006: 42 – 43) argues, the 'political room for a Western-style environmental movement still seems limited, as international NGOs themselves have found'. Indeed, *Greenpeace* and the *World Wildlife Fund* have invested major efforts in further stimulating the environmental movement in China, with mixed success.

Rather than consisting of a united mass of people confronting specific power groups, the environmental movement in China is, on the whole, fragmented, highly localised, and non-confrontational. This is due to various factors. As Peter Ho and Richard Edmond (2007: 331) argue, contrary to popular belief, China's burgeoning civil society is not state-led, but rather one in which state – society relations 'is not a matter of the former dictating the latter, but rather a kind of "negotiated symbiosis"' in which environmental activism resourcefully adapts to, rather than opposes, the political conditions of its era. As a result, China's social activists have developed 'a diffuse, and informal rather than formal, network of relations', which has regardless 'yielded undeniable political as well as societal legitimacy' (Ho and Edmond 2007: 331). This specific setting has caused environmental activists in China to be enmeshed in a 'diffuse web of informal relations, unwritten rules, and shared missions with the party-state' (Ho and Edmond 2007: 334).

Both the use of online environmental activism to foster a sense of individuality, and the conception of Chinese environmentalism as fragmented and localised, rather suggests that environmentalism in China cannot be called a 'movement' as such. Unlike the beginnings of environmental movements in Eastern European countries, which were often tied to the desire for regime change, manifested in a non-overtly political form, and therefore highly homogenous, such beginnings in China were not linked to wider appeals for democracy or human rights (despite many Western commentators being keen to interpret them as such). However, as Doyle and McEachern (2001) argue, there is no unifying teleological purpose that drives social movements; no single causal reality that made them; and interpretations of the origins and significance of environmental movements are as contested as the movements themselves. Considered in this light, and as a 'depoliticized politics' and 'self-imposed censorship', as Ho and Edmond (2007) argue, used strategically and self-consciously to foster its own existence, the collection of environmental concerns and action in China can most definitely be seen as a movement – and one that young people are instrumental factors in supporting.

Roots and Shoots, Global Village, Friends of Nature and other NGOs tended (and indeed still do tend) to particularly attract young people, as was their intention. Therefore, the environmental movement in China, as in many countries, was, from the outset, a youth movement essentially. Although older people are also of course concerned with the environment, especially when it directly affects them, it is young people who have taken up the cause in the greatest numbers and with the most energy.[4] This concern on the part of young Chinese is certainly primarily about the environmental issues themselves, but is also part of a new self-expression that is particularly valued amongst China's younger generation in contrast to the more collective attitudes of their parents' generation. Young people are increasingly part of a generation for which the highest aim is the ability to steer one's own life course as an individual. This makes being passionate about 'causes' such as the environment, a way to show one's personal beliefs and passions, that are, crucially, personal to oneself (even if they are shared with millions of others). They are about identity formation and staking a claim for who one is and what one believes as an individual (see Sima and Pugsley 2010).

Although such concerns can be seen to form a common thread across both rural and urban young Chinese people, it is the young urban dwellers who are most likely to be ensconced in this relatively new culture of individualism. The growth of the xiaokang, mentioned above, is particularly fast in Chinese cities; it is here that the emphasis on making statements through one's actions, and often through one's consumptive choices, is strongest. New, young members of the xiaokang form the majority of environmental campaigners, and they are, almost by definition, urban. The environmental movement, then, can be seen as part of a new urban, individualistic culture, that creates subjects that Tang Xiaobing would see as the height of 'post-revolutionary' in nature.

Chinese environmentalism and ChinaNet

The internet arrived in China in 1994 and grew in coverage and usage incredibly quickly. Statistics from the China Internet Network Information Center (CNNIC) show that there were approximately 10,000 internet users in 1994, whereas by December 1998 the number of internet users had reached 2.1 million, and by December 2002, 59.1 million. It was, and still is, most heavily used by young people with relatively high levels of education, and is far more prevalent in urban areas. Amongst more educated urban young people 38 per cent of Chinese bloggers were students, 51 per cent had college education and above (a figure much higher than the average education level of Chinese netizens) and that China's blogosphere is still largely populated by those rich in 'cultural capital' – students and elites (CNNIC 2007).

The onset of ChinaNet usage was contemporaneous with the burgeoning environmental movement in China. In fact, *Friends of Nature*, the first and one of the most influential environmental NGOs in China, was also founded in 1994 – the same year China was connected to the Internet. Therefore, it is no surprise that ChinaNet began to be used in conjunction with other channels in order to encourage people to become involved in environmental NGOs, and later, campaigns and protests. In short, environmental concerns, urban youth culture, and the internet were an unavoidable collision. Despite this, according to Yang Guobin (Yang 2003a: 89) the role of the internet in the environmental movement has been overlooked. Little has been written about the huge increase in the numbers of websites created not only by green NGOs, but also by government agencies and research centres, and individuals interested in specific environmental topics ranging from green lifestyles to the protection of endangered species. According to Yang, based on his study of four typical NGOs, there are three key ways in which ChinaNet plays a role in China's environmental movement; first, it enables voluntary environmental activity with minimal financial resources in a restrictive political climate. NGOs frequently rely on volunteers with their own personal internet use who can help gather information, edit material, and undertake website design and maintenance. This has the added advantage of meaning they do not need office space. Second, an online presence means organisations do not need to register at a specific geographic location, therefore avoiding regulations that state there can only be one organisation for any specific type of work, in any given location (e.g. Beijing could only have one NGO working on endangered species). Finally, an online presence provides a key sign of a group's existence – a factor that is as important to the volunteers as it is to the outside world and that enables NGOs to showcase their work at a community level by using their websites to showcase their offline activities. This raises environmental consciousness, promotes environmental discourses, and helps mobilise the public – sometimes spurring public action or gaining coverage in national newspapers and TV programmes (Yang 2003a: 90 – 91).

Importantly, ChinaNet has facilitated the move offline for many environmental groups that began their existence online only. In this way, internet technology is facilitating the creation of new organisations for social change, and provides a space for ordinary citizens to organise and act collectively in a manner that Yang argues is genuinely bottom-up (Yang 2003a: 91). It is Yang's contention that 'over time the green Web that Chinese environmentalists are weaving may reach far and wide' (2003a: 91). Natalie Wong also points to the crucial role played by ChinaNet in the environmental movement in China. Wong argues that despite the tightening of laws and the increase in policies to monitor use, environmental activists continue to use ChinaNet as a key means to organise activities. Campaigns frequently largely rely on online forums to recruit volunteers and discuss their plans. Such forums support information exchange and advocacy, and provide opportunities for the public to engage in discussions. Monitoring and suppression on the part of the authorities has simply made activists more tactical in how they post their messages (Wong 2015: 154).

For many commentators ChinaNet is also key in enabling the younger generation in China to live out the new-found self-expressionism mentioned previously in this chapter. For example, Yangzi Sima and Peter Pugsley (2010: 287) argue that China's 'Generation Y' are more affluent and better educated than their parents, often the only child in the family,[5] and the first to grow up in an internet world. This combination causes them to consider individuality a highly sought-after quality, and has given rise to a 'me culture' (*ziwo wenhua*) primarily concerned with self-expression and identity exhibition, often through mass entertainment and consumerism. For Sima and Pugsley (2010: 287) this decidedly post-socialist mentality of individual expression, achievement and pleasure has taken over from the arguably 'collective interest' mentality that marked the older, Mao generation, and can be seen as being lived out online. Their argument is, in part at least, based on a study of blogs, which, they argue, increasingly contain bloggers' 'personal confessions' (47 per cent), record personal daily activities (41 per cent), and provide a platform where people discuss their interests and hobbies (31 per cent) (figures from CNNIC 2007: 18). As Sima and Pugsley (2010: 288 – 289) argue, personal blogs, therefore, clearly dominate the Chinese blogosphere in terms of number and popularity, and despite their seemingly banal, exhibitionist content, have an impact frequently underestimated by journalists and researchers in the West. In China, in terms of blog readership, 43 per cent of people surveyed by the China Internet Network Information Centre (CNNIC) said they often visited blogs about people's personal experiences, and 42 per cent visited blogs where authors wrote about their daily activities (CNNIC 2007: 32).

Such self-expression is clearly a feature for the current generation of young people in China and a factor that the internet, and indeed ChinaNet, has always lent itself to. It is therefore, perhaps not surprising that the online world has become a conduit for asserting individuality, but that collective concerns, such as those over the environment, can be part of this assertion. Recent protests, such as the PX[6] protests that have taken place since 2007, are testament to the growing role of ChinaNet when it comes to the environmental movement and indeed the organisation of offline protests, especially in cities. Indeed Chin-Fu Hung (2013) argues that the case of the Xiamen PX movement was a key part of the evolving phenomenon of internet-empowered environmental activism and proves the enhanced public participation and environmental rights defences that are developing in China. For Hung (2013: 40), this is providing a formidable challenge to the Communist Party of China and has a socio-political impact that will be shaped not by the technology itself, but by 'the underlying political dynamics of public opinion, civil participation, citizen journalism and cyber-activism'.

Starting with a pioneering local campaign against a multibillion yuan PX petrochemical plant in the eastern coastal city of Xiamen in 2007, China has witnessed a series of large-scale protests against PX plants and other large-scale industrial and infrastructure projects in recent years. The initial Xiamen campaign occurred when the National Development and Reform Commission formally endorsed plans to build a PX plant near a new residential area in Haicang, a suburb of Xiamen, despite earlier rulings by the Fujian Provincial Environmental Protection Bureau and the State Environmental Protection Administration to relocate the plant further away from residential areas. Having tried, and failed, to communicate with the government through personal networks, a Xiamen university professor then called for the relocation of the plant on the grounds that the project was too close to residential areas and posed a high accident risk for the wider public. Despite the motion not being adopted, the professor managed to obtain support from 105 government delegates, including the deputy governor of Fujian Province. This gained substantial national news media coverage and motivated more local advocates to become involved in the campaign, calling on line and via mobile phones for on-the-street protests. Updates on China's social media platforms and via text messages used code words and memes in order to escape detection from the authorities and make their way through China's firewall. The protest gained identity by the use of a distinctive yellow banner, and discursive elements such as the idea of the people of Xiamen inhabiting an intangible and collectively owned 'beautiful Xiamen' (*meili de Xiamen*) that was under threat and thus required citizens to 'defend the beautiful home' (*baowei meili jiayuan*). According to Steinhardt and Wu (2015) this message became viral online and was present throughout the protests.

In addition, times and places to meet in person were communicated to key supporters without the authorities knowing, and these supporters then passed on the information to others – creating a network of supporters largely undetectable to official powers. As a result of this strategic online campaign and organisation, the first *offline* protests took place on 1 and 2 June 2007 with several thousand participants in each 'collectively strolling'[7] though Xiamen city centre.[8] Individual demonstrators' accounts and photographic evidence reveal widespread sympathy from large crowds of bystanders in the streets of Xiamen, and there were reports at the time of acts of sympathy from police officers. Many of the participants had reservations about taking to the streets, fearing they would be hurt by police action, or arrested, yet support for the cause was widespread. In response, the State Environmental Protection Administration conducted another environmental impact assessment of the PX project, followed by a public hearing, confirming in December 2007 that they would not go ahead with the project.

What was most interesting about the Xiamen PX protests however, was the extent to which they linked diverse local people together in a distinct movement that placed particular emphasis on the urban as a beautiful and collective environment. As Steinhardt and Wu (2015: 34) point out, 'protesters claimed to speak not for a narrow subset of citizens, but for the general public of this major city' … 'demonstrators' accounts and photographic evidence reveal an outpouring of sympathy from large crowds of by-standers and even occasional acts of sympathy from police officers on the days of the protests'. Even members of the Xiamen diaspora elsewhere in China showed support online before and after the protests. Indeed, the Xiamen protests were so distinct that they triggered 'copycat protests' in other large cities also due to have PX plants built. Online and then on-the-street protests followed in the cities of Dalian, Shifang, Ningbo, Kunming and Chengdu (and quite possibly many other cities that have escaped the attention of Western media sources due to Chinese censorship). This meant that young people from cities across China felt linked and united and were creating new online solidarities with each other despite their often very different backgrounds and lifestyles. As Steinhardt and Wu (2015: 35) argue the PX protests in Xiamen were a 'transformative event', as 'longer-term structural changes provided the opportunity to innovate contentious strategies and *significantly change the popular imagination of what a protest can look like and achieve*' (my italics). It proved that social elites and average citizens can join forces against a major state-backed project and be successful in preventing it going ahead. Moreover, it created a sense of a common urban culture amongst the (mainly) young people involved in the protests. As one anonymous young person was reported as saying during the Xiamen protests, 'We don't want GDP. We just want to live.'

In addition, the protest was post-materialist in the sense that those who took part often did so not for issues that would affect them and those they knew, but for their fellow citizens and because they believed *in principle* it was wrong for the

PX plants to be built. This was not the NIMBY-ism that Thomas Johnson speaks of (see Johnson 2010), and broke with the form of collective resistance focusing on everyday grievances that Steinberg and Wu see as coming to the fore in the early 1990s and dominating for the two decades that followed. Rather, it was part of a new protest repertoire that is preventative, rather than seeking compensation for 'victims' post facto; that is witnessing the gap between those who protest and those who engage in policy advocacy narrow substantially; and crucially, that revolves around public goods, policy concerns, and symbolic values relevant to a broad 'imagined community' of a city or even the whole country (Steinhardt and Wu 2015). It is also important to note that the emphasis here is on the city as the site of this imagined community – the new repertoire of protest is specifically urban. Wasserstrom (2009) too speculates that protests indicate a rising participatory demand among the middle class and signify the emergence of a fledgling version of ecological modernisation in China. A modernisation in which the presence of ChinaNet is instrumental.

Environmentalism, ChinaNet and civil society

Since the introduction of the internet into China in 1994, a lively internet culture has emerged. Online chatrooms run by Netease.com attract tens of thousands of users at almost any given time of the day; the number of netbars flourishing throughout the country continues to increase at a dramatic pace (and this despite government efforts to keep them under control), especially in cities and urban areas. ChinaNet has triggered public debate over a wide variety of issues, as well as shaping social organisations by expanding old principles of association, facilitating the activities of existing organisations and creating new associational forms, thus prompting many commentators to champion its role in the creation of a public sphere in China. Many commentators have also suggested that the creation of such a sphere may be a step towards democracy in China although Guobin Yang (2009), a key commentator on the progression of the internet in China, is less certain about its potential on civil society in China. Yang (2009) argues that Chinese online activism derives its methods and vitality from multiple and intersecting forces, and state efforts to constrain it have only led to more creative acts of subversion. However, despite wholeheartedly acknowledging that the internet in China is enabling users to organise, protest, and influence public opinion in unprecedented ways, through a range of contentious forms and practices, Yang (2003b: 453) argues that political power and market forces are insufficient conditions, and whilst they may have given birth to the internet as a new sphere of social and political life, they also take away this life, as indeed they did in Habermas's analysis of the eventual refeudalisation of the public sphere. He points to internal and external obstacles that hinder democratic participation in online communication on ChinaNet. Internal obstacles are factors such as cyber 'bullies' who attempt to deprive fellow participants of their voice; the time and practice required to being capable of meaningful discussion, articulating ideas, taking positions, inquiring into others' positions, etc.; stylistic and rhetorical indecencies that make online discourse uncivil. These are, according to Yang (2003b: 473), not insurmountable, unlike external obstacles that are economic and political in nature, such as government control over content, the commercialisation of content that leaves little space for political action, and the fact that only the better off in China have access to ChinaNet. This said, Yang (2003b: 474 – 475) acknowledges that ChinaNet has given rise to a new type of political action in China – online critical debate, through which 'citizens are becoming better informed about and more engaged in social and political affairs'; collective protest has been enabled by ChinaNet; and in addition, existing social organisations have developed an online presence and virtual communities have been built, especially by environmental groups, in ways that link to global civil society. The nature of this collective protest is often one of constant micro-critiques, not dissimilar to Elizabeth Perry's (2008) point about twenty-first-century protest in China being 'rebellious' as opposed to 'revolutionary', but that such constant small rebellions can still lead (and indeed have) led to substantive social change.

Arguably, such critique taps into the desire amongst China's young, urban generation to prove their individuality by picking the causes they believe in and being vocal about their beliefs. It can be seen as part of what Lisa Rofel (2007: 197) describes as Chinese citizens' search for a 'new cosmopolitan humanity' that has emerged due to what she describes as the upheavals within the uncertainties of social life in contemporary China. Crucially, this new humanity is self-aware; concerned as much with how China is seen by the world as it is with its own subjectivity. Indeed, Zhang (2008: 21) argues that notions of identity creation through blogs are precisely 'an extension of Chinese desires to inhabit this new global space' and provide a narrative discourse of modern Chinese subjectivity – 'identity, selfhood, interiority, and self-image (or rather self-imaging)'. Being part of critique, then, is also part of carving out a new subjectivity, and of recognising that subjectivity as part of a global awareness. To be an environmental activist is (potentially, and often), to be part of a young, urban elite who are globally self-aware and recognise themselves as part of a global environmental movement. Indeed, many NGOs play upon and promote this global awareness to the young urban Chinese they hope to attract to their cause. As Johnson *et al.* (2007: 361) argue, *Roots and Shoots* adopt practices and foci that are relevant on a local level, but are equally concerned with facilitating connections among young urban people regionally, nationally, and globally, in order to promote social justice. Arthur Mol (2006: 30 – 31) too, argues that globalisation itself meant that a 'global civil society, global environmental governance, and environmental management systems operated by transnational corporations in developed and developing countries' are no longer restricted to Western countries.

Conclusion

China's young urban environmentalists are fighting not only for causes they believe passionately in, but also for the ability to carve out their own individual identities, as part of the constantly negotiated relationship with the Chinese state, and as members of a new type of public sphere that reflects this constant critique and aligns itself to a modernity based on a global outlook. As such, they confirm the arguments that environmental struggles, across global locations, defy the post-materialist rationale and are about far more than achieving security for future generations. Many of course have argued that this was never in doubt, but the case of China brings it to light in particularly pertinent ways. China's young, urban environmentalists are part of a generation fighting not only for environmental justice, but for the ability to live out their own lives in ways that enable them to feel in control of their individuality and their position as subjects in the current form of modernity in China. It is a battle for the boundaries of cultural existence in the city, as much as it is for the saving of the environment.

Perhaps most pertinently for a discussion about Chinese urbanism, it is also a battle that challenges current thinking which tends to dwell heavily on a new generation of urban Chinese citizens concerned with property ownership, wealth accumulation, and the status symbols of global brands. This newly enriched urban xiaokang tend to be characterised as highly pragmatic (even socially irresponsible), consumerist, individualistic, and concerned with China's continued economic development at all costs (including environmental) – the 'social person' mentioned previously is essentially seen as a straightforwardly self-interested person. Considering urbanism and urban cultures via the urban environmental movement forces a reconsideration of these elements and a more nuanced approach. The xiaokang can be considered as desiring continued improvement in living standards but not at any cost; as seeking ways of reconciling consumer desires with responsible and sustainable development; and as seeing and living individualism not as a series of selfish acts, but as *choosing* one's own allegiances, as opposed to historic community allegiances.

Notes

1 ChinaNet is the name given to China's Internet due to the way in which the firewall used by the Chinese government to control the content available on it effectively makes it a national internet as opposed to a global one.

2 The term Xiakang, despite being ancient, was reintroduced by Deng Xiao-ping in 1979, and has remained a key term and goal for the Communist Party of China ever since.

3 Roots and Shoots is an environmental and humanitarian program for young people formed by Jane Goodall, the world-renowned chimpanzee researcher and United Nations Messenger for Peace. Dr. Goodall first established Roots and Shoots in Tanzania, East Africa, in the early 1990s. Today, more than 8,000 groups exist in almost 100 countries. Roots and Shoots' mission is 'to foster respect and compassion for all things, to promote understanding of all cultures and beliefs, and to inspire each individual to take action to make the world a better place for humans, animals and the environment' (Jane Goodall Institute 2003).

4 See Weiya Huo's article: https://www.chinadialogue.net/article/show/single/en/1549-The-new-face-of-youth-activism-in-China.

5 Due to the one child rule brought in by Deng Xiao-ping in 1978, many families in China have, until recently, only been allowed to have one child. There were always various exceptions, including the rule not applying to ethnic minority families. The rule began to be formally phased out in 2015.

6 PX stand for paraxylene - an oil-based chemical product that is widely used in plastic bottles, polyester fibre, etc.
7 Collective strolling (*jiti sanbu*), is a term used to avoid the word protest and legitimize the action in the eyes of the state, as protest is illegal, but walking together is not.
8 Figures and reporting from Didi Kirsten Tatlow and Joey Liu, 'Thousands Protest against Chemical Plant in Xiamen', *South China Morning Post*, February 2, 2007, http://www.scmp.com/article/595260/thousands-protest-against-chemical-plant-xiamen, accessed June 19, 2014.

7

INTERSTITIAL SPACES OF CARING AND COMMUNITY

Commodification, modernisation and the dislocations of everyday practice within Beijing's *hutong* neighbourhoods

Melissa Y. Rock

Introduction

Yuanqin buru jinlin (远亲不如近邻) is a Chinese idiom that means 'close neighbours are better than distant relatives' and underscores the vital role geographic proximity can play in the creation of strong community relations and social networks. While family relationships are important bonds that can contain multiple and overlapping investments into the physical, mental and social health and wellbeing of family members, physical distance can impede the ease at which the access and distribution of family welfare is made available to 'distant relatives'. Thus, 'close neighbours' become key substitutes for relatives who are too remote to rely on for certain small, but crucial daily practices – such as watching over the elderly or young for short periods of time, keeping eyes on the street to assist neighbours in need, observing both the mundane and extraordinary activities that traverse the alley to identify and stand guard against intruders or perceptible danger.[1]

It is argued that in the transition towards a socialist market economy, China has newly redrawn literal and figurative boundaries of public and private space in favour of the private, whereas previously during the Mao era, the reverse had been valorized (see Chapter 2, this volume). In the early post-Mao era, the spaces of public and private in old Beijing's alleyway (*hutong*) neighbourhoods dynamically overlapped and evolved. As such, these interstitial spaces often served as unique locations for rich social interface, community care and civic engagement, especially for the more vulnerable segments of society (i.e. the poor, young, elderly, disabled). During the early post-Mao economic reform era, families and neighbours residing in Beijing's center city hutong have continued to live largely in close quarters – fluctuating between harmony and disharmony – creating and maintaining tightly interwoven relationships that are often mutually beneficial, if not socio-economically necessary. However, processes of commodification, modernisation and gentrification expedited at the turn of the century, especially in preparation for the hosting of the 2008 Summer Olympic Games, have notably worked to squeeze out and displace interstitial spaces (Brighenti 2016). These contemporary urban processes are underpinned by China's unique historical trajectory beginning in the Mao era, continuing through the reform era and beyond.

Drawing from numerous China scholars (Broudehoux 2004; Wu *et al.* 2007; Zhang and Ong 2008; He and Wu 2009) who closely examine China's particular engagement with neoliberalism, in this chapter I think through the ways in which China's transformation into a socialist market economy shape and frame the rich tapestry of social relationships, or encounters in Beijing's hutong neighbourhoods. I also consider the concomitant disruption to and rebounding of public and private spaces and the spaces in between them that this transformation brings. Geographers He and Wu (2009) posit that China's 'market-driven socio-spatial transformation in the realm of urban redevelopment' – its 'emerging neoliberal urbanism' – highlights the role that market forces take in shaping urban landscapes. Their analysis places Chinese cities at the forefront of capital accumulation strategies, rationalising a far-reaching assortment of socio-economic and spatial reconfigurations. Building from this emphasis on China's particular engagement with neoliberal practices in urban centres, I investigate how the changing market forces are applied to (the privatisation of) residential housing and incorporated into the logic of everyday socio-spatial practice. Thus, through an examination of Beijing's hutong neighbourhoods, a history on which I elaborate upon in the next section, I suggest we can better appreciate the ways in which economic restructuring has prompted drastic change in *siheyuan* (courtyard housing) residential ownership, occupancy and demographic makeup.

China's state sponsored neoliberal engagements have not only worked to produce 'modern' cities, but have also been utilised to cultivate 'modern' citizens. Although engagement with neoliberalism restructures the relationship between capital and the state, forays into neoliberalism need not imply a weakening of state control over citizen subjects, according to anthropologists Zhang and Ong (2008). Rather, neoliberal rationality provides a technology of rule that induces 'citizens to be self-responsible, self-enterprising, and self-governing subjects' (Zhang and Ong 2008). They argue that efforts to mobilise the citizen's capacity for self-governance are at the heart of the neoliberal reasoning that guides privatisation. As this chapter demonstrates, China's emerging neoliberal urbanism catalyses a restructuring of state – society relations that plays out in the domain of public/private/interstitial living space(s) and the social reproductive practices contained therein.

While rapid urbanisation continues to facilitate increasing geographic proximity and population density, neoliberal commodification and privatisation combined with discourses of personal responsibility (Zhang and Ong 2008), modernity (Rofel 1992; Hershatter 2004; Hanser 2005) and embodied quality, or *suzhi* (Anagnost 2004), have effectively succeeded in eroding the overlapping public-private interstitial spaces that have fostered and fomented intimate neighbourhood encounters and interactions. Further, the restructured state – society relationship plays out in (or is indicated through) transformed material and sensorial encounters that occurred throughout Beijing's hutong neighbourhoods. Indeed, the dislocation of everyday practices from the interstitial spaces of old Beijing marks a splintering and fragmenting of a social fabric that notably serves and supports its most marginalised and vulnerable residents (e.g. female care givers, young and elderly people, poor and disabled). This chapter elucidates the ways in which Beijing's economic and spatial reforms frame and constrain socio-spatial practices of caregiving and community engagement.

Reconfiguring residential space in Old Beijing

> [P]lace meanings are inves(n)ted by people ... they are constructions – selective accounts – combining current and daily experiences and distilled memories ... places are closely tied to community, spatially anchored or symbolically constructed ... that identity is derived from the tangible and non-tangible ... that there is a politics, broadly defined, in the investment of meanings in places, whereby the powerful often have the wherewithal to define places and their meanings ... the meanings also reflect the process of negotiation
>
> *(Yeoh 1996: 10)*

The quote above illuminates the fluid meaning of place and its reciprocal role in formulating notions of community and identity. The reconfiguration of residential space can entail the modification to the physical structures, the rearrangement of public, private and interstitial space, or the various transgressive uses and repurposing of the space by occupants and visitors. Changes in the spatial orientations, their subsequently derived meanings and the everyday uses, of *hutong* neighbourhoods, reflect processes of residential adaptation (or resistance) to the economic restructuring ongoing in contemporary China. The residential history of old Beijing hutong neighbourhoods is one that connects dynamic swings in socio-political and economic foundations of the state with the nationalisation or privatisation of land (and housing) and the subsequent consequences to the organisation of social reproduction and everyday practice. Thus, siheyuan and their hutong neighbourhoods have distinctive constructed meanings throughout various periods in their history contingent upon social and economic structure as well as its political and legal designation.

Piper Gaubatz (1995, 1999) details three distinct eras of city formation: the pre-1949 city, the Maoist city (1949 – 1978) and the emerging Chinese city. The defining character of the pre-1949 city was the functional differentiation and specialisation based on clan or place of origin relationships between residents and/ or occupational specialisations (Gaubatz 1999). All major architecture was aligned with the cardinal directions, grand city walls demarcated city limits, and the entire city (except for monuments) was built low to the ground in one- and two-storey structures (including siheyuan, or four-sided courtyard houses), the height and extent of ornamentation of which were dictated according to resident status within a social hierarchy. Courtyard residences traditionally housed just one family. During this era one's courtyard house reflected relative rank, affluence and influence in society. However, due to long-term neglect and severe overcrowding, by 1949 the once-elegant urban patterns of the imperial era had significantly deteriorated.

In an effort to create a decentralised and self-sufficient urban society, the Mao government seized the opportunity to institute socialist-type planning and organization by (1) establishing cities as industrial development production centers, and (2) encouraging work units (*danwei*) to become self-sufficient communities within the city. In addition to employment, a danwei provided for housing, health care, food distribution and other social services (Gaubatz 1999). Supplanting the function of the city wall, each danwei enclosed its work-unit compound within a walled boundary under the assumption that most urban residents had little need to travel outside of their work/neighbourhood sphere.

However, the structural formation of traditional hutong neighbourhoods in old Beijing restricted industrial development in these neighbourhoods during the Mao era. As a result, most hutong remained primarily residential. Some work units acquired courtyard houses to subdivide and distribute among their employees, while others were subdivided and utilised as neighbourhood committee social service provision centres. Thus, during the Mao era courtyard houses were redistributed such that families lived in tighter spatial proximity, with multiple families now occupying one siheyuan – each family residing in one of the four quadrangle buildings around the courtyard. The post-Mao era (1979 – present) has experienced equally dramatic changes with respect to the planning and organisation of the city and lived spaces. The role of the work-unit has diminished significantly with regard to providing housing, health care, education and other social services (Bray 2005). By the late 1970s, Beijing experienced a housing shortage as newlyweds looked for their own housing, a large influx of people returned to the city after years of rural re-education during the Cultural Revolution, and the population of employment-seeking rural-to-urban migration grew rapidly (Davis 1993). As a result, subdivisions of courtyard houses continued as their inner courtyards were filled in with additional occupants and amenities. Migrant workers often built 'temporary' hutong housing, adding onto and reinforcing these structures over time (see Chapter 13, this volume). Overcrowding in many of the *hutongs* led to their disrepair and designation as city slums.

China's transition to a socialist market economy has facilitated the creation of commodity housing and a robust real estate industry. Both national legislation and local regulations have facilitated the development of urban housing markets (Wang 2000). In the 1980s, property titles of some collectivised hutong housing units were reclaimed by the previous owners or relatives of the former owners – if proof of prior ownership could be produced (Kaiman 2012). In other cases, many publicly owned houses were sold to existing tenants or other public sector employees (Wang 2000). This patchwork of public and private ownership of siheyuan housing created a complex mix of interest and investment in both the physical and social structure of the courtyard houses and their hutong neighbourhoods. Some families who were offered the opportunity to regain private ownership of their courtyard house declined because the responsibility for maintenance and repair would shift to their own shoulders. Considering the poor condition of such units after decades of neglect, many of the poorest and more elderly residents were ill-equipped to take on such a task. In choosing to remain public housing tenants, they relied upon the local authorities to maintenance and repair their homes when needed. Those families which did accept the offer to regain ownership of historically held property rights still had to contend and negotiate with the current tenants living there – some having occupied the residence for decades. For example, in negotiating public-to-private residential ownership and occupancy one former hutong resident, Mr. Zhai, now in his mid-forties, explained that he had been raised from early childhood in a courtyard house sub-unit. One of the other families in the courtyard house had historically owned the place before Chairman Mao had come to power. After it was divided and collectivised, Zhai's family was among those moved into the courtyard space. The family had regained property rights to the place in the 1990s and wanted Zhai's family to move out. He refused and remained in the sub-unit in which he was raised. In 2003, the family offered him 130,000 rmb to move. Zhai refused, and remained put. Finally, in 2007 they offered him 380,000 rmb. He accepted this offer and moved into a rented low-income hutong housing unit only a few blocks away.

In addition to complex private ownership arrangements, some courtyard house units comprise both public and privately owned units or a *pingfang* (one-storey units that compose one side of a three or four-sided courtyard house unit). Walking through old Beijing, visitors will notice numerous courtyard house entryways with both a "公" (*gong*, or public) *and* a "私" (*si*, or private) insignia at the top of the main doorway. The units inside are therefore a mixture of both public and privately owned one-storey configurations – making renovation, rental and sale negotiations, not to mention social relationships quite complicated, strained and replete with challenges. Courtyard houses which have multiple private owners must negotiate these numerous challenges and conflicts of interest as each family jockeys for the best financial position and potential gain. The privatisation of Beijing's siheyuan alters the socio-spatial terrain upon which community networks are formulated and social reproductive practice unfolds. Thus, recuperation of hutong property rights in the 1980s and continuing today, remains exceedingly complex and fraught with social, spatial, and economic tensions and implications.

Spaces of social reproduction

Spatial proximity is one element that facilitates neighbourliness and communal care efforts. But in and of itself proximity is not sufficient for the promotion of mutual aid. The functional quality and formation of residential spatiality also lends itself to various configurations of communal care and social reproduction. Contemporary Chinese cities have adopted a new spatial logic of residential ownership and occupancy in response to neoliberal economic reformations that reorganise the socio-spatial practices of everyday life.

China's transition to a socialist market economy marks a purposeful shift in the Chinese state-market-society relations. As the state pushed to grow their economy, Chinese citizen-subjects were called upon to participate in state-prioritised market endeavors no longer as model labourers (working 'to serve the people'), but as consumer-subjects with middle-class aspirations, increasing purchasing power, and an emergent consumer appetite (Zhang and Ong 2008). They were incentivised and disciplined to accept this new logic of organisation through a concentrated rhetoric centring on consumer choice (*xuanze*), individual quality (*suzhi*), and the necessity of becoming 'civilised' (*wenming*) modern consumer-subjects (Rofel 2007; Zhang and Ong 2008; Zhang 2010). This new organising logic has re-bordered and re-ordered the intimate social and spatial residential formations within Beijing's hutong alleyways. After siheyuan housing was collectivised in the 1950s, social and spatial conventions of engagement changed dramatically. The domain of the 'public' (*wai*, or outer) via the state (or Chinese Communist Party, CCP), entered and usurped the 'private' (*nei*, or inner), exposing all concepts of 'the private' to state scrutiny (see Chapters 2 and 6, this volume). A vast academic and historical literature recounts the manner in which these events took place and shows how, among the resulting impacts of these tumultuous campaigns and reforms, socio-spatial formations and relationships were reconfigured around an integration with and dependency on the social institutions embedded within the state apparatus (such as work-units, pensions, health care, and community residence committees). Thus, the spaces of the public and private, where the 'order of the home was understood to be deeply interconnected to the order of the state' (Goodman and Larson 2005), consequentially trouble western binaries of strict public/private domains. The residential spaces of a transitional Beijing exemplify the spaces through which these twentieth-century reconfigurations materialised – as the national and city governments slowly and increasingly transitioned towards a socialist market economy. In concert with the economic transition, the mutability of Beijing's in-between spaces began to waver as a public/private (and *nei/wai*) spatial dichotomy began to re-inscribe itself back onto the landscape.

The rich social life in many of Beijing's hutong neighbourhoods is both a characteristic and consequence of living in close quarters – first mandated by the collectivisation efforts in the 1950s and 1960s then subsequently sustained through the conditions of urban poverty as vulnerable populations (i.e. the poorly educated, elderly, and mentally or physically challenged) were slow to adapt to the quick transition towards state capitalism. One of the legacies of the Mao era was the consolidation of residential space during collectivisation. As additional families were moved into newly nationalised siheyuan, individual family spaces shrunk overall. Further, as various functions of social reproduction were collectivised, such as food preparation and child care, spaces of production and social reproduction began to merge. Public and private spaces and their respective everyday practices became increasingly interwoven. Official policy, such as '(t)he "five transformations"' (*wuhua*), for example, called for the 'socialization of housework, the collectivization of daily life, the universalization of education, the normalization of hygiene, and the greening of communes' (Bray 2005). The scope of public space and activity enveloped and usurped that of private spaces and activities.

In the post-Mao era of market-based economic reforms, spaces of production and social reproduction began to untangle. In particular, residential housing, once subsidised by the central government has increasingly transitioned to commodity housing, such that by 2010 the rate of home-ownership in China had reached 80 per cent (Barth *et al.* 2012). As the complex residential ownership landscape within Beijing's hutong neighbourhoods further materialised and carved out newly privatised spaces for family re-appropriation, communal use value and practice has increasingly been shut out and in search of new spaces for engagement and activity. However, the market shift from subsidised housing to commodity housing has not instantly charted new, hardened boundaries for socio-spatial caregiving and community practices. Rather, the in-between spaces of community and caregiving continue to spill out into public spaces, or enter into private spaces, as residents negotiate the new socio-spatial logic being hammered out alongside China's contemporary socio-economic reconfigurations.

Interstitial spaces of caring and community

During China's transition to a socialist market economy, public and private spaces and the everyday practices contained therein continued to intertwine and unravel in multifaceted ways. Owing to the relatively high population density in Beijing's hutong neighbourhoods and siheyuan, where multiple families often live in tight geographic proximity, private life has a tendency to spill over into public spaces. In warmer months it is not uncommon to walk through Beijing's hutong neighbourhoods and observe laundry drying on clothes-lines just over the threshold of a siheyuan and extending into the interstitial alleyway spaces (see Figure 7.1).

In the heat of the summer, some hutong residents may wash their hair and brush their teeth just outside the doorway of their siheyuan – performing private practices in these interstitial spaces for lack of convenient or appropriate space to do so within their residence. Another notable practice in Beijing's hutong neighbourhoods includes an evening stroll through the alleyways in comfortable, button-down pajamas (Penêda 2012). Beijing's hutong have functioned similarly to Jane Jacob's (1961) sidewalks in Greenwich Village: as places that facilitate contact and socialisation among residents, but most importantly for children and the elderly. These mundane everyday practices conducted in Beijing's public, private, and the interstitial spaces in between them highlight an unspoken intimacy between neighbours. Such intimacy can draw neighbours closer together as they become the backdrop of each other's lives.

For example, in the early 2000s, Mrs. Zhang lived in a hutong neighbourhood just southeast of Tiananmen Square when her mother was still alive, but in poor health. Having resided in Beijing's hutong neighbourhoods their entire lives, they had become accustomed to negotiating the particular challenges put forth by ancient-style architecture that was not particularly accommodating to

FIGURE 7.1 Clothes drying on a makeshift clothes-line in an alleyway, Beijing

Source: Melissa Yang Rock

the constraints of Grandmother Zhang's wheelchair mobility. As such, the elevated door in the entranceway to their siheyuan posed one obstacle to overcome before leaving the inner threshold of their home. Though inconvenient, the Zhang family was able to lean on their neighbours for assistance. When her mother was ready to venture out, Mrs. Zhang opened the doorway to her siheyuan and announced, 'Lao da ma is going out! Come (help us move her)'. Neighbours gathered to help lift Grandmother Zhang's wheelchair over the elevated doorway and past the step-down onto the paved alley. Parked near her courtyard home, she sat, chatted, and observed the neighbourhood around her. Neighbours came and went about their daily business, making sure to engage with and keep an eye on Grandmother Zhang. 'Would you like some more water, lao da ma?' people asked. Food? A blanket? Sun? Shade? Neighbours and long-time workers at local businesses (often residents from nearby neighbourhoods) would check up on Grandmother Zhang while her daughter stole away the spare moments to run errands, prepare dinner, or wash clothes. Mrs Zhang explained that in past decades, when her mother was healthy, she was a vital member of the community – and looked out for the health, safety and wellbeing of their neighbours. In return, when Grandmother Zhang's health began to decline and her family felt the increasing strain placed upon them not just by an ailing elder, but by a government that had slowly begun its transition towards a socialist market economy, catalysing a retraction from provisions of social services, neighbours did what they could to help ease that burden. Mrs Zhang recounted how neighbours would show up bearing extra fruits or vegetables they had bought

FIGURE 7.2 Carving out their own social space, courtyard residents sit socialising on a couch placed in a *hutong* outside their *siheyuan* entryway, Beijing

Source: Melissa Yang Rock

at the store, explaining how the neighbour knew that lao da ma enjoyed eating fresh dates or lychee.

Relationships built up over years spent socialising with neighbours in these hutong alleyway spaces created not just an elevated level of mutual care for each other, but also a duty and responsibility for each other's health and wellbeing. If Mrs Zhang had not been able to rely on her neighbours for caretaking assistance, she would have been challenged to finish her various chores and run errands. Further, her mother would have had limited access to the very dynamic social sphere ongoing in hutong neighbourhood spaces – to the richness and texture that can give meaning to life and living, especially when faced with declining health and limited physical mobility.

This is but one example of the ways in which private life spills out into public space, creating new in-between spaces of caring and community. The communal intimacy forged in these spaces is borne out of daily encounters as well as the mundane practices of social reproduction – such as through exercising on the outdoor public gym equipment (see Figure 7.3) within the hutong.

This intimacy does not happen without tensions (see Chapter 15, this volume) – as the intimacy I highlight does not assume harmony – but it does connote recognition, familiarity, and curiosity even when expressed through airing of conflict and objection. Beijing's hutong neighbourhoods are experiencing a demographic

FIGURE 7.3 Public exercise equipment in a Beijing *hutong*. One of many publicly available exercise stations placed throughout the city

Source: Melissa Yang Rock

change as many siheyuan have transitioned to private single-family residences and commodities, while others remain a patchwork of public housing (danwei) and rental properties. As such, siheyuan owners have applied a number of strategies in order to capitalise on their asset. While siheyuan are somewhat standard in design (a quadrangle formation), the partitioning of space over the past few decades combined with historic variation based on the socio-economic class of the original occupants, has created a very dynamic market for the value of hutong living. As a result, the 'rarified' hutong and siheyuan have become places that embody the tensions between the push for modernisation, while attempting to capitalise on a nostalgia for the traditional old Beijing hutong landscape.

Conflicts arise as property values increase, catalysing swift upward pressure on rents, leaving residents either priced out or compelled to move. The delicate cohesion of the mixed income hutong neighbourhood is further undercut by modernisation discourses that identify and isolate various behaviours and practices occurring within interstitial spaces mentioned above as low-quality (*suzhi dì*) or backwards (*luohou*). Often, in these cases applying such discourses works to further stigmatise and castigate the poor and migrant populations living in the hutong. These discourses echo market-based pressures on hutong neighbourhoods that discourage the use of interstitial and public spaces for 'low-quality' or 'backward' practices – influencing the retreat of these practices back into private spaces, or the dislocation of those individuals who are prone to practise them. Such modernising discourses work to further support the continued commodification and gentrification of Beijing's hutong neighbourhoods.

Private cars in public alleyways

In addition to discursive pressures to re-inscribe a more distinct nei/wai socio-spatial orientation, the modern demands placed upon Beijing's hutong neighbourhood infrastructure tend to exacerbate the tensions over social reproductive use of interstitial spaces. Chief among these is the desire for private car ownership. The rate of car ownership in China has skyrocketed over recent decades, growing at a rate of 37.4 per,cent annually since 1985 (Zhu *et al.* 2012). The demand for private car ownership brings with it the demand for parking (Collins and Erickson 2011). As such, Beijing's hutong neighbourhoods face great pressure to reconcile modern demands of private car ownership with its ancient city design and the constrained alleyway spaces therein. Beijing's historic hutong, typically ranging between 3 to 9 metres wide, were not built to support the large-scale movement of automobile traffic. Designed primarily for pedestrian use, hutong alleyways have been ill-equipped to accommodate Beijing's rising car traffic and associated parking requirements. Nonetheless, despite these spatial constraints, car ownership and its associated parking spaces are in high demand within old Beijing's hutong neighbourhoods (see Figure 7.4), just as it is throughout urban China (Lu 2016).

FIGURE 7.4 Cars parked in a *hutong* alleyway, Beijing

Source: Melissa Yang Rock

Parking is not permitted in hutong alleyways narrower than 6 metres. It is permitted in hutong greater than 6 metres, but only in spots designated by lines painted on the roadway (Shoup and Jiang 2016). These regulations have not prevented the widespread practice of illegal parking. Rather, the ubiquitous nature of this practice overwhelms the parking authorities to such a degree that they largely tolerate much of the illegal parking, usually issuing tickets only for the severe obstruction of an alleyway (Shoup and Jiang 2016). As such, private car owners put forth great effort to commandeer, occupy and protect their hutong parking space – whether legal or illegal.

In some cases, formal barriers are provided to secure reserved parking spaces, while in other instances, owners have taken to placing informal barriers such as locked bikes (see Figure 7.5) or cinderblocks in the way to deter others from parking in the elusive hutong parking spot. Siheyuan owners who have renovated their homes to include a garage proclaim the space immediately in front of the garage doors to be 'no parking' zones to ensure that they maintain access to movement in and out of their garage (see Figure 7.6). In addition to these tactics for preserving reserved legal parking spots, illegal parking further overwhelms hutong neighbourhoods, effectively privatising public space to store a personal commodity. This encroachment into the interstitial alleyway spaces serves to push out the multi-generational everyday practices still enjoyed in smaller alleyways (Figure 7.7) where cars are prevented from entering (although smaller three-wheeled electric vehicles, *sanlunche*, do manage to commandeer some of that space, see Figure 7.8). Children, the elderly and disabled are discouraged from playing next to the cars for lack of space; but also for safety reasons, as smaller children are not easily visible darting in between cars, and as the elderly and disabled are slow to move.

Forgivable trespasses: private spaces and everyday practice

Just as there exist interstitial (or 'in-between') spaces that have blended private life and everyday practice into public spaces, the private spaces of local residents have also been encroached upon. Interstitial spaces thus far have been characterised as private life that spills over into public space. However, the transition from collective households to commodity housing did not instantly render the all the spaces of siheyuan private. In mixed public/private siheyuan complexes, the red entrance gate was often left open allowing the free flow of residents and guests in and out of the courtyard housing. As a result, neighbourhood intimacy and spontaneous community interaction can temporarily extend interstitial spaces into the interior of siheyuan.

Such was the case for Li Laoshi (Teacher Li) who reflected upon the concepts of private spaces and boundaries in the hutong by recalling one particular neighbour who would often make her way, following her nose, into Li Laoshi's kitchen before dinnertime (see Chapter 9, this volume). Using the obvious ice-breaker of food preparation to make her entrance, the neighbour would inquire about the contents of the eventual meal before moving on to other topics of conversation. The

FIGURE 7.5 Bicycles locked to fixtures on the side of the road to protect *hutong* parking spaces, Beijing

Source: Melissa Yang Rock

FIGURE 7.6 Hand painted sign warning people not to park in front of the car garage, Beijing

Source: Melissa Yang Rock

FIGURE 7.7 Foot traffic and an elderly man sitting on a *dengzi* (small stool) in a narrow *hutong* interchange, Beijing

Source: Melissa Yang Rock

FIGURE 7.8 The many private uses of a narrow *hutong* alleyway include parking bicycles, three-wheeled vehicles (*sanlunche*), drying laundry on a clothes line, and other miscellaneous storage, Beijing

Source: Melissa Yang Rock

boundary of the doorway was an inconsequential indicator of 'privacy'. Rather, this symbolic territorial marker had been remade into an in-between place, one neither exclusive nor fully private, it was a meeting place for neighbours (and not a marker of privacy against one's neighbours). In this context, to deny this elderly neighbour the joy and utility of access to Li Laoshi's kitchen would have been rude to the long-time neighbour. Although Li Laoshi sometimes wished she could have a 'private' place, she also valued the intimacy with her neighbour. In recounting the story, Li Laoshi seemed more amused than annoyed upon remembering her neighbour's antics. Because, for all the inconveniences of hutong living (such as poor sanitation, lack of private bathroom facilities, deteriorating building structures, and too many people and cars), Li Laoshi still missed the richness of her interactions and life in the neighbourhood that marked her first fifty years. Forging intimate relationships within Beijing's hutong neighbourhoods is not only predicated upon direct encounters between people in the hutong, but also inclusive of the sensorial aspects of the sights, sounds and smells enjoyed (or endured) by being present and observing the material and immaterial aspects of place. Li Laoshi's account of her neighbourly interaction underscores the numerous ways people are brought together to create memories, as well as attach meaning and value to socio-spatial practice rooted in place.

In Peter Hessler's (2006) reflection in *The New Yorker*, he underscores the ways in which sounds permeate through walls – transcending boundaries that contain material objects but only slow or muffle audible noise (see Chapter 15, this volume). His account reveals the residents' rhythmic connection (through sound) to each other, as well as to the vendors who transect their neighbourhoods selling essential staples, random wares and seasonal goods. These sensorial encounters begin in the early morning, without even leaving the confines of Hessler's room.

> Usually I'm awake by dawn, and from my desk I hear residents chatting as they make their way to the public toilet next to my building, chamber pots in hand. By midmorning, the venders are out. They pedal through the alley on three-wheeled carts, each announcing his product with a trademark cry. The beer woman is the loudest, singing out again and again, *"Maaaaiiiii pii-iiijiuuuuuu!"* At eight in the morning, it can be distracting – "Buuuuyyyy beeeeeeeeeer!" – but over the years I've learned to appreciate the music in the calls. The rice man's refrain is higher-pitched; the vinegar dealer occupies the lower registers. The knife sharpener provides percussion – a steady click-clack of metal plates. The sounds are soothing, a reminder that even if I never left my doorway again life would be sustainable, albeit imbalanced. I would have cooking oil, soy sauce, and certain vegetables and fruit in season. In winter, I could buy strings of garlic. A vender of toilet paper would pedal through every day. There would be no shortage of coal. Occasionally, I could eat candied crab apple.
>
> *(Hessler 2006: 12)*

Encounters with the sounds of the hutong are daily and varied. In addition to hearing vendors hawk their wares from their three-wheeled bicycles (sanlunche), residents may also hear the street sweepers tidying the alleyway, a neighbour's radio playing next door, or the cooing of pigeons from another resident's coop. In 2009, in an offshoot hutong between *Nanluoguxiang hutong* and the Drum and Bell Towers, I lived across from a family raising rooftop pigeons. Every summer evening my neighbour would release the group of one hundred or so pigeons to fly freely. They would circle the sky above our neighbourhood, stretching their wings to soar above old Beijing. A whooshing sound swirled overhead as the birds flew by repeatedly, en masse. In a city of over seventeen million residents it marked a moment in the day when earth-bound residents' attention lifted skyward.

Discourses of modernisation in conjunction with the processes of commodification and gentrification enshrine new valuations and sensibilities within hutong neighbourhoods and their residents. New spatial partitioning in Beijing's hutong suggests that modern living is bounded, contained and guarded. Further, new social codes reflect the desire for individual privacy and independent living, and increasingly dictate reconfigured social formations and relationships within Beijing's urban residential spaces. But the preferences for modern notions of privacy are not universally shared. While some residents find nosy neighbours to be intrusive and unwelcome, others may welcome the company and chance to banter. Similarly, the intrusion of a vendor's distinctive call into the interior spaces of the home might inconvenience or perturb some. However, others are comforted by the predictability and accessibility that the vendor's call represents. Women who relied upon neighbourly 'eyes on the street' to help with communal care activities (of young and old) feel the increasing strain of care work fall upon their individual shoulders as everyday practices retreat into private (domestic) spaces. Lastly, the elderly and infirm tend to be more dependent upon, and therefore more nostalgic for, the blurred spaces and messy encounters of everyday life in the hutong.

Don't visit me: We need the privacy (bie canguan wo, women xuyao yinsi 别参观我：我们需要隐私*)*

The old city rarified ***hutong*** neighbourhoods are relentlessly marketed as containing 'authentic' Beijing culture (also see Chapter 10, this volume). When tourists, both foreign and domestic, visit China's capital city, one stop sure to be on their sight-seeing list is a journey through one of old Beijing's trendy hutong neighbourhoods. *Qianmen Dajie, Nanluoguxiang* and *Shichahai* are but a few of the recently commercialised hutong drawing in large numbers of visitors on a daily basis. As these hutong increase in popularity, the commercial opportunity in the nearby neighbourhoods has risen markedly. Over a short period of time, cafés, restaurants, as well as boutique clothing and accessory shops have popped up throughout the residential hutong spaces. Drawn to these alleyways in search for an 'authentic' Beijing cultural experience, tourists walk through them not just to appreciate siheyuan architectural construction, but to observe how their everyday residents live in them. The physical intrusion into siheyuan residential spaces by tourists near these popular hutong hot spots is common enough to have prompted some residents to post signs in Mandarin and English requesting that passersby respect their privacy and refrain from entering their private home (see Figures 7.9–7.11). The quest by outsiders for an authentic hutong experience has forced some local residents to mark their private territory, close their front gates, and draw their daily interactions back to the interior of their courtyard.

In search of 'authentic' experiences in the 'disappearing' hutong

The desire for the authentic Beijing cultural experience has created a market demand that is met through various types of hutong and siheyuan tours. These tours provide an architectural and cultural history of old Beijing's courtyard houses and traditional neighbourhoods, often as the patron sits in a pedicab-driven rickshaw. One stop along the tour is an actual courtyard home. Depending on the tour company organising the outing, the group may experience a more grand-style courtyard that has been renovated to reflect its historical form and function. Or they may be brought to a more modest home and given a chance to see a

FIGURE 7.9 Sign posted on courtyard door asking passersby not to visit and to respect their need for privacy, Beijing

Source: Melissa Yang Rock

FIGURE 7.10 Hand written sign posted on courtyard door explaining that the courtyard is 'closed to visitors, please do not disturb', Beijing, China

Source and translation: Melissa Yang Rock

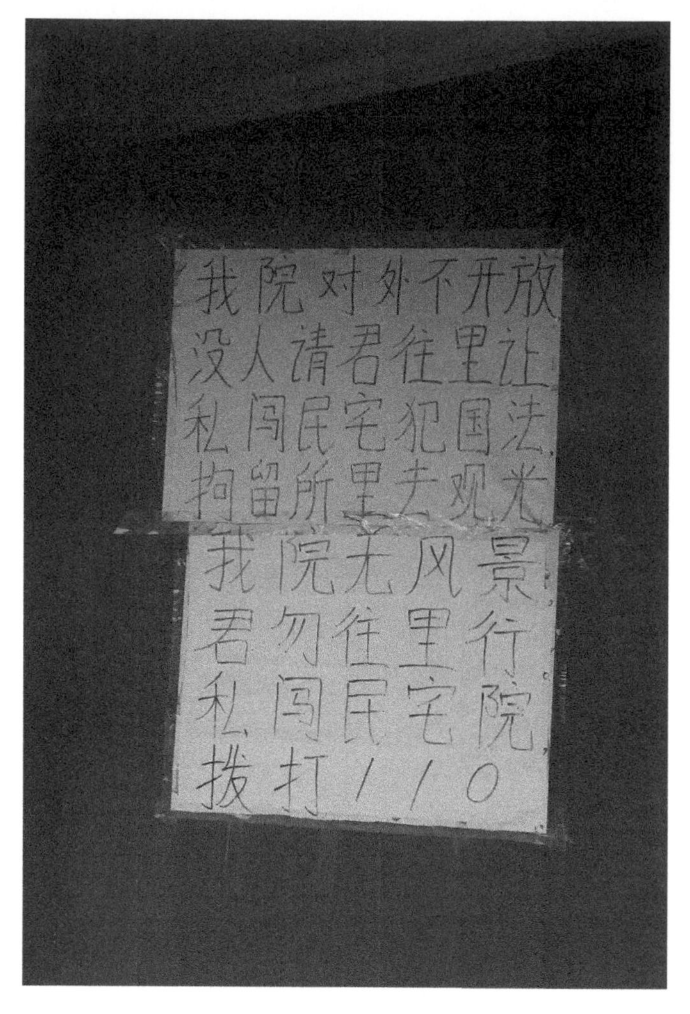

FIGURE 7.11 Sign posted on courtyard door, Beijing, that states:

Our courtyard is not for outsiders' admission,
No one invited you in or gave permission.
Trespassing into people's homes breaks the law,
And then you'll be sightseeing in detention.

Our courtyard is not scenic,
You're forbidden to come in it.
For trespassers on our courtyard,
We dial 110 for the cops.

Source: Melissa Yang Rock. Translation: Kristine Harris

humble living arrangement reflective of the Mao-era communal living arrangements. Whatever the experience, money changes hands from tourist to tour guide to owner or resident offering a peek into the interior lives (even if contrived or performed) and spaces of hutong residents.

People in search of a more extended hutong experience may avail themselves of short or long-term siheyuan rentals. A quick search on peer-to-peer or list-and-rent home sharing sites like Airbnb.com, or a local Chinese equivalent such as Tuijia.com or Xiaozhu.com, highlights numerous short-term overnight options for international and domestic tourists alike to experience a taste of hutong living. While some residents may offer up a room in their own home to the sharing economy, others will renovate portions or all of their siheyuan for dedicated short-term rental. This trend not only potentially removes these properties from the long-term rental housing market, putting upward pressure on rental rates, it also elicits some trepidation by hosts and their neighbours over the increase of strangers passing by and through their living spaces. This has created a new market for a more active property management of these peer-to-peer rented units (Tsang and Mozur 2017). Managers can be effective at vetting guests, responding to their needs, and addressing any breaches in house rules and decorum. Siheyuan and hutong guests treat these rentals as their home away from home. But as they do, interstitial spaces disappear. Instead, private space is carved out (of the interstitial space) for Airbnb-type customers while local residents' everyday practices retreat behind the confines of closed doors. The social fabric in the alleyways further frays as the greater percentage of outsiders living temporarily within these neighbourhood spaces dilutes the frequency and intensity of contact among permanent neighbours whose daily rituals in turn deviate from their regular patterns in their attempt to moderate interaction with siheyuan strangers and guests.

Conclusion

In old Beijing's hutong neighbourhoods, the idiom 'close neighbours are better than distant relatives' continues to underscore the vital importance of geographic proximity for the creation of strong community relations and social networks. However, the spatial proximity that facilitates neighbourliness and communal care efforts relies upon interstitial spaces to provide opportunities for societal interaction, community investment and social reproduction. While the cases laid out in this chapter are not necessarily unique to Beijing's hutong neighbourhoods, the process by which they are normalised, articulated and solidified through discourses centered on modernity, management and the regulation of public and private spaces (Mitchell 2003) – as well as the interstitial spaces 'in-between' them – highlight the ways in which socio-economic and spatial access are governed by China's emerging neoliberal urbanism (He and Wu 2009). Nonetheless, there is resistance to the disciplining policies and discourses that seek to appropriate, displace and dislocate interstitial spaces of caring and community. Here I do not refer to organised resistance, but instead a focus on everyday practice that continues to push back on the privatizing forces placed on interstitial spaces.

As such, the changing geographies of care in Chinese cities warrant critical inquiry into who is differentially impacted by the shifting spatiality of care work and social reproductive spaces and practices. In this regard, literature on the changing geographies of care under neoliberalism (Massey 2004; Mitchell 2004), and feminist scholarship on China's socio-political and temporal fluctuations as they affect gendered spatial practice (Gilmartin 1994; Goodman and Larson 2005) aid a more nuanced understanding of the changing gendered divisions of care work in contemporary China. The privatisation of residential space experienced during China's reform period has exacerbated a re-feminisation of caregiving responsibilities in Beijing's new urban spatiality (Rock 2012). Thus, spatial dislocations that force care work from public and interstitial spaces back into the private sphere result in the redistribution of caregiving back primarily into the domestic (female) domain. The unequal gendered division of care work stems from re-inscribed notions of public/private socio-spatial dichotomy that shape how gendered identities are (re)constituted and performed within the confines of Beijing's gentrifying hutong neighbourhoods.

This chapter explores the ways in which everyday practice is transformed within hutong neighbourhoods to accommodate the spatial logics and discourses associated with Beijing's neoliberal urbanism (He and Wu 2009). I demonstrate that although the socio-spatial restructuring experienced in contemporary hutong neighbourhoods provides its residents with a high degree of geographically partitioned proximity, processes of commodification and gentrification have eroded the interstitial spaces where *close* or intimate social encounters build tight, mutually beneficial social networks. Rather, the commodification of Beijing's courtyard housing (siheyuan) and concomitant propagation of discourses of modernity and personal quality effectively create, reinforce and normalise distinctive boundaries that demarcate not just private ownership or usury rights, but also the social conceptions of responsibility within a neoliberal socio-spatial sphere of everyday interaction. Thus, the informal structures of social reproduction which had previously been located in the interstitial spaces within hutong neighbourhoods are squeezed out of the newly re-inscribed contemporary binaries of spatial organisation and social practice in Beijing's new commodity housing framework. The long-standing saying, 'close neighbours are better than distant relatives', once an indication of Beijing social life, is less descriptive of life in its neoliberal urban present.

Note

1 This chapter draws on data gathered through 11 months of ethnographic fieldwork in Beijing, China conducted in 2009 and the summer of 2015. Methods include informal interviews with *hutong* residents, transect walks, and visual analyses of *hutong* neighbourhood spatial layout, use and signage in select old Beijing neighbourhoods. The names presented in the case studies below have been changed to protect the identity of research participants.

PART III
Consumption and urban cultures

8

TASTING, SAVOURING, SIGNALLING

Articulating the luxury brand experience in Chinese cities

Annamma Joy, John F. Sherry, Jr. and Jeff Jianfeng Wang

Introduction

The gossamer texture of a Hermès silk scarf, the buttery, supple leather of a Louis Vuitton lambskin handbag, have, along with other luxury goods from Europe and North America, become valued in urban China. Western emblems of refined taste, whether Cartier jewels, Ermenegildo Zegna suits, Rolex watches, or Gucci shoes, have become *de rigueur* among Chinese individuals seeking to signal social status, in contexts both professional and personal. We argue that over recent decades, with the accelerated incorporation of Western ideals of status and self-reward by both wealthy and aspirational consumers, a new luxury brand experience, one unique to Chinese culture, has taken root.

Initially, political bureaucrats and employees of multinational corporations comprised the urban Chinese elite; today, government officials, business managers, private entrepreneurs, and other professionals have expanded that cohort (Liu-Farrer 2016: 501). As of this writing, China is the world's largest consumer of luxury brand goods (Chang 2014); Chinese purchases comprise almost half of all luxury brand consumption worldwide (*China Daily* 2016). To explore this developing luxury brand consumption experience, we travelled to Beijing, the People's Republic of China's capital city; Shanghai, the unofficial fashion capital of China; Nanjing, a rapidly growing commercial hub, and Hong Kong. We conducted interviews with and observed our participants (drawn from the ranks of various luxury brand stakeholders, including consumers, fashion professionals, salespersons, and manufacturers), during our multi-year ethnographic study.

Our chapter is grounded in the 'politically and ideologically mediated and socially contested' nature of contemporary cities that is a focus of critical urban theory (Brenner 2009: 198). The 'urbanization of the world' (Brenner 2009: 199) is accomplished in no small measure by the inexorable spread of consumer culture, and the processes of accommodation and resistance by which this global force is given a local shape. The 'intensification and extension of the urban process' (Brenner 2009: 205) is reflected in 'brandscaping', by which consumers construct material and symbolic environments with marketplace products, images, and messages – an investment in local meaning, whose totemic significance largely shapes the adaptation individuals make to the contemporary world (Sherry 1995, 1998, 2005). Murakami-Wood and Ball's conception of brandscaping (2013) embodies the fundamental characteristics of emplacement and experience design, but is more carceral and managerial in scope.

Local adaptations of brandscapes, or 'consumptionscapes' (Venkataraman and Nelson 2008), occur when urban Chinese consumers adopt iconic brands (such as Starbucks) through on-site enactments of personally meaningful experiences, transforming the space to suit their own needs (Venkataraman and Nelson 2008). We explore the practices and conditions, such as embodiment (Jayne and Leung 2014) that underlie the luxury consumption experience in urban China, to understand how Western brands contribute to the redefinition of urban social relations. More specifically our research explores the practices of tasting, savouring, signalling through which luxury brands are embraced as emblems of status and refinement. Additionally, we identify the temporal, spatial, and social components that constitute experiential consumption of Western luxury brands. With little history of heritage luxury branding, Chinese consumers view Western luxury as inherently desirable, evoking a future unmoored to a more restrictive past. We explore urban Chinese consumers' desire for style and modernity exemplified by Western brands, and for accumulating professional, social and cultural capital through luxury consumption, as practised by the affluent, nouveau riche, and aspirational Chinese upper-middle class. By assessing the prevailing view of luxury brands as a variant of contemporary art, we further provide a thorough understanding of the complexities and nuances inherent in luxury brand consumption in China.

Articulating the luxury brand experience

Knowing, becoming, and doing are key aspects of experiential luxury. We use the terms 'tasting, savouring, signalling' to capture the essence of urban Chinese consumers' luxury brand experience. We argue that tasting and savouring, far from being private experiences, are rather completely social and learned – particularly so in China, given its long cultural history with Western luxury. Signalling, as an acquired social construction perceived by both consumer and observer, is by definition reliant on shared values. In their experience of luxury brands, our participants focused on aesthetic aspects, highlighting the importance of experiential knowledge in becoming a luxury connoisseur, with time a central factor. Note that our Chinese participants were often the first generation in their culture to experience luxury brands, and thus focused on the intense pleasure – immediate and palpable – that they experienced as first-order sensations, captured in the term 'tasting'. First-order and second-order sensations help consumers prioritise their preferences and create a sense of urgency to consume. First-order sensations were overtly apparent at social gatherings that we attended in Hong Kong, marked by 'oohs' and 'ahs' from appreciative friends as they assessed each other's respective branded looks. Lightning-fast judgements differentiated connoisseurs from novices depending on whether their brand use was judicious or indiscriminate. As one participant noted, 'For those who are knowledgeable, less is more. Simplicity and elegance are key …'.

Participants who gave significant attention to 'savouring' worked in media or the arts and as managers of luxury brands, or simply had extensive experience with luxury brands. We consider them taste-makers (De Certeau *et al.* 1998), because they serve as fashion educators for their avid followers and customers, passing on the narratives surrounding a brand, including history and heritage. They themselves are somewhat detached from the typically intense sensory experiences associated with consumption. Such detachment is not unusual, since art experiences have similarly been described as second-order experiences (Csikszentmihalyi and Robinson 1990). The Chinese concept of savouring (*pin wei*) reflects a particular approach to living (Frijda and Sundararajan 2007). We normally associate savouring with a pleasant experience, in which the individual wishes to prolong the moment because of heightened pleasure. In the act of savouring, the object being savoured is not the ultimate focus; instead, the contemplation and meaning associated with the object are central. Individuals are able to detach themselves from the object of consumption and show restraint in their responses.

'Signalling' refers to how participants broadcast status. Luxury goods are tools to demonstrate their owners' economic, social and cultural capital (Bourdieu 1987). Costly, classically elegant luxury brands denote both obvious wealth and refinement. Such cultural capital, indicative of savouring, reflects a deep knowledge and appreciation of luxury goods, often linked to art and high fashion (see Chapter 10, this volume). The ability to revel in product quality, design, colour, texture and tailoring reflects personal style and *savoir-faire*. By contrast, social capital needs to be gained through the approval of others, as observers confirm economic capital, which in turn leads to cultural capital, and thence to social capital. All three types of capital – economic, cultural, and social – are directly acquired via luxury goods consumption.

A temporal-spatial orientation of the luxury brand experience

We build on the temporal-spatial dimension of consumption (Arnould and Price 1993; Russell and Levy 2012; Thompson 1997; Tumbat and Belk 2010), and the implicit notion of time past, time present, and the future. Our argument rests on Throop's (2003: 233 – 234) variations of temporal orientation. In addition to the Husserlian notion of retention (looking back) and protention (looking to the future), Throop documents at least four different types of temporal orientation; each may differentially structure the experience of the self and the world. These include an orientation to the present moment consisting of unfulfilled protentions as open anticipation of an indeterminate future; an explicit future orientation comprising imaginal anticipations of a determinate future predicated on residues of past experience; a retrospective glance that entails the plotting of beginnings, middles, and ends over the already elapsed span of a delimited field of experience; and the subjunctive casting of possible futures, even possible pasts, across the fluid space between past and future.

An explicit future orientation based on past experience.

Many of our participants focused on both the luxury brand and the internal self during the savouring process. Kim, an editor with a Beijing-based fashion magazine, noted that: 'There are lots of people with money, but they have no taste'. Liu-Farrer (2016: 506) refers to this state as *tuhao* (gaudy nouveau riche), and argues that education is required to create an elite consumer culture. To gain such expertise, Kim deliberately studied her stylish supervisor: 'I noticed how she moved her body, how refined she was. There was deliberation in what she did and said. That is a skill learned only with time.' In Kim's comments the plasticity of the body and the importance of self-fashioning are paramount (Joy and Venkatesh 1994; Thompson and Hirschman 1995). As Kim recalled: 'I once asked a Louis Vuitton salesperson: "Is this [handbag] a new model? It looks like an old edition." He said that it was new, although to an untrained eye it may look older: "There are differences. The colour of the lining is lighter, and the zip is of a different yellow." I was amazed to discover such differences, and to know that the salesperson was very knowledgeable about them.' As Kim reveals, tasting and savouring are distinct. Basking in the pride associated with ownership is only one aspect; being the recipient of attentive service, as she was, is equally gratifying. Russell and Levy (2011) refer to such experiential savouring as reconstructive re-consumption, which shapes a positive evaluation of the experience and thereby encourages the customer's return.

Kim's retrospective glance at an in-store experience allows her to plot a narrative, as Throop (2003: 234) describes, with beginnings, middles and ends over the elapsed span of a delimited field of experience. Her narrative would bring her back to the store repeatedly: a determinate future. The object of Kim's pleasure exists, but is enhanced by her own experience. Knowing gives her greater pleasure, which she sustains by social engagement with the salesperson to learn the differences in models from one season to another, or between authentic and counterfeit products. The relational aspect of pleasure is important. Her range of emotions involves both the immediate sense of joy at the discovery of special features, as well as the extensive appraisal of the events generated through her contact with the salesperson, which suggest that savouring puts into play the temporal self – the self associated with re-tasting (*hui wei*), in contrast to the self of tasting in the present (Russell and Levy (2012)'s reflexive re-consumption). In the process, Kim signals her own understanding of luxury culture and, through its prism, her own social status. As exemplified by Kim's comments, initially a consumer experiences the recognition of sensation. Later, Kim provides a cause for her feelings – the salesperson's attention to detail – and then ends with narratives that are not divorced from the body. The relational nature of the experience with the salesperson is also critical to her self-reflexivity. Savouring is the most prominent experience, although tasting and signalling are taken for granted in her narrative.

Skill is not merely the embodiment of knowledge, but also of the physical, perceptual and behavioural change that accompanies enskillment (Downey 2010; Joy and Sherry 2003). For Kim, Louis Vuitton and France are constructed as 'other', a significant factor that promotes her appreciation of this brand. She also attributes a lack of taste among consumers to a Socialist past, contrasting 'China' with 'France'. Dong and Tian (2009: 517) remind us that Western brand consumption and the creation of subjective meanings is not routine apolitical behaviour. Kim's position approaches what Dong and Tian (2009) identify as a view of Western brands as a means to achieving higher social status.

Another participant, Brenda, works in a luxury store in Shanghai, and is a tastemaker who educates her customers about luxury brands. Inter-subjectivity is apparent in her discussion of her role as a personal advisor to wealthy clients:

> The brands that I like do not have factories in China. It is important to have the real thing ... Ninety percent of my jewellery is from Cartier. I know what I am getting: quality and beauty. My shoes are Prada ... I hate large logos, including LV. I do not need to show the logo in order to be happy and confident ... [and] feel very comfortable ... I show people how to put together a look that is polished, refined, and harmonious ... [my customers are] keen to learn. You cannot look down on them; you have to bring them up to your level.

Brenda is discerning about what brands suit her lifestyle, evidencing a second-order evaluation of a hedonically based first-order sensation. Brenda's distaste for products manufactured in China is a phenomenon much noted by luxury brands (Pang 2008). Some well-known luxury brands do not admit that components are made in China or other less developed countries (Reinach 2005), since the final stages will be completed in France or Italy. The experience of artisanal rarity, far removed from exploitative labor practices associated with cheaper brands, is key to the luxury experience (Kapferer 2010). China is constructed in Brenda's mind as a 'place' that has poorer standards than France and Italy and is the home of fake products. Brenda believes that a sense of heritage will happen in time, even as there is no certainty that her clients will become refined. Brenda is optimistic; for her, the value of changing political ideologies (trading the remnants of the former Socialist approach for a cosmopolitan view) has resonance to dreams of wealth (Dong and Tian 2009). Like Kim, she makes only passing references to current political ideology, because her focus is on the future. Her temporal orientation could be defined by Throop (2003: 234) as 'an explicit future orientation that consists of imaginal anticipations of a determinate future predicated on past experience'.

Both Kim's and Brenda's descriptions embody order and coherence, in which savouring underscores the agentic aspect of the self. One hears in their reflections the sense of 'completeness' that Thompson (1990: 355) suggests is important for the experience of being in control. They exhibit a concern for a temporal process accelerated through knowledge and learning, although their own experiences belie the importance of acceleration. As Rafferty (2011: 252) notes, 'Women learn to relate to self-fashioning practice – a practice of "choosing" – in disparate ways, depending on their class position at origin, their related trajectory development (experiences), and the emotional capital bestowed to them.' Ironically, the consumer values the deference displayed by salespeople, while superior taste requires that the salesperson suspend judgement and bring the client up to her level.

Marlene, a wealthy homemaker from Nanjing, deliberately cultivates knowledge:

> I spend a lot of time learning about brands … There are many brands in Italy that interest me, like Blue Marine … wearing this brand is better than wearing a brand with a well-known logo. It [a logo] is no longer what a girl longs for, and I like the feeling when I wear clothes that make me look good. It is a question of harmony. I even move differently when I wear a Blue Marine dress or skirt. I don't want to worry about how other people view me, as long as I like how I look …

Marlene's comments suggest a learning curve for knowledge about luxury brands. The clothes become you, she suggests, reflecting an acquired emotional habitus. They are so embodied that they condition physical movement. She also values her own judgement, thereby signalling confidence and status. The benchmark of being knowledgeable is important. Not everyone knows the Blue Marine brand, whereas everyone knows Louis Vuitton. Within the boundaries established by luxury brands, Marlene creates her own style. She embraces discretion: a set of embodied practices that conceal and reveal potentially significant information, and that performatively establish a subject's positionality within a specific community of practice. Discretion calls attention to a contextualised set of revealing and concealing practices. By the logic of discretion, objects can radiate beyond their material limits to suggest a thoroughness of interpretation of those 'in the know'. Symbols are ubiquitous, but only the correctly conjured person has the necessary knowledge to decipher, to participate, and to see (Mahmood 2011: 431). Marlene notes that her clothes have to suit the occasion, recognising that diversity must be balanced with social regulation. Giving face to others is more important in this context than saving face. Savouring to her is a self-initiated action of attention and absorption mediated with concern for others. Throop (2003: 234) would describe Marlene's temporal orientation as future oriented with imaginal anticipations of a determinate future, in common with the other participants discussed above.

For Kim, Brenda and Marlene, refinement is an ongoing process, whether personal or related to the goods they consume. Illouz (2009: 401 – 402) argues that luxury goods exemplify the inherent instability associated with refinement. Refinement represents the process of infinite gradations of objects, with the newer versions being better and more nuanced than the earlier. In turn, when such goods are consumed they embody refined taste. The differences among Marlene's unevenness of learning and inconsistency, Kim and Brenda's unhurried pace, and the piecemeal acquisitions of all three, suggest that mimesis is complex. Like Üstüner and Holt (2009), we observe cultural diversity (not a unified habitus) in the ways in which each of these women relates to luxury.

A present orientation with a dose of pragmatism

Joan, a lawyer in Nanjing, is discerning in her choice of luxury brands, with the assistance of professional stylists:

> I buy Chanel, Prada, Hermès, and Louboutin. The first brand I bought was a Prada bag, because I needed some spiritual support in being a lawyer. I am not a twenty-year old. I do not like large logos ... [or] things that are shiny and fancy. They are for younger people. Many people may buy the same brand, but how you put it together is the mark of how different you are from others. Experienced people will figure out with one glance what you are wearing. I feel good when people praise me for my good taste.

Joan noted that her Prada bag covered her body, giving her confidence in her interactions. Handbags, although limited in their respective sizes, loom large for Joan, providing a sense of protection. Her comments suggest that other than subjectification (the objects enter the self: Belk 1988) and materialisation (you enter the object: Lastovicka and Sirianni 2011), objects can also provide a shield, literally and metaphorically, echoing the authors' argument that consumers use objects to fill an interpersonal void. Joan orchestrates a dynamic equilibrium of difference and diversity, complementing Japanese brands with Western brands, and finding balance and harmony in knowing that she looks elegant and has the approval of others. Signalling social status is important to her. Savouring entails prolonged processing to better appreciate the experience. Joan decided to learn about luxury brands later in her life; she quickly acquired knowledge, but could not entirely shake off her lack of confidence. Embodied habitus takes time; Joan is more present oriented, but hopes in the future to be comfortable in her own skin.

For our participant Lacy, an accountant in Hong Kong, brand clothing is essential to her professional life:

> When I take clients to dinner, I [wear luxury brands] because [otherwise] I feel inadequate. This is why I use a Chanel (for exclusive contracts), not an LV bag. When you carry the right bag, the bag becomes the star of the evening. It may be ... big or small ..., but it covers your body like a flowing robe. I don't feel inadequate anymore. The client appreciates my knowledge

> of luxury brands, and is willing to do business with me. I end up confident and successful in making deals.

Lacy reveals a strategic use of luxury goods, with her bags serving as both protective wraps and sources of confidence. Strikingly, Lacy uses the same image as Joan: that a luxury handbag, whatever its size, metaphorically covers the body. As a talisman, the bag confirms its owner as a member of the affluent elite, and alleviates feelings of inadequacy (Lastovicka and Sirianni 2011). Her self-presentation is elegant and assured; therefore, she herself is as well. What boosts Lacy's confidence is neither solely her skill nor her professionalism, but also her choice of luxury handbag. Her choice to wear Chanel for special business contacts matches its relative exclusivity compared to Louis Vuitton. Both clothes and bag are a second skin, secreting powerful emotions and enabling dramatic performances. Luxury brands also reaffirm her beliefs about their essential role in her personal life: 'I pair LV bags with jeans when I go out shopping. It is casual but elegant and always an asset. Otherwise salespeople treat you with condescension until you are ready to make a purchase.' We viewed such treatment firsthand in visiting luxury brand stores in Mainland China, where luxury shopping is less overtly ritualised, and consumers are often aspirational; sales representatives were seen treating cash-paying customers (as opposed to credit-card paying, favoured by the affluent) with thinly veiled disdain. In Hong Kong, where consumers tend to be truly wealthy, customers were treated with courtesy. Many Chinese thoroughly enjoy buying genuine luxury goods under such circumstances. Brand representatives initiate them into the details of the brand, package purchases artfully, and return proffered credit cards with both hands in a sign of respect, resulting in an experience that lingers and begs for repetition, and is intrinsically linked to the brand itself. In both her professional and personal lives, Lacy feels protected from disdain or other negative experiences simply by carrying a specific handbag. Lacy has a future orientation based on present and past experience (Throop 2003: 234). Joan and Lacy exhibit status anxiety and both rely on the market to express their authentic selves. While each of our participants above experiences similarities in temporal and emotional processes, their respective life trajectories and social classes differentiate their experiences.

A future orientation but with emergent, imagined and critical anticipations

Jenny, a fashion consultant in Beijing, spoke of a 2007 Fendi fashion show, reporting that the designer Karl Lagerfeld's inspiration came from an old Chinese painting at the British Museum; Lagerfeld particularly liked the cheongsam (a traditional, form-fitting long dress):

> I find the clothes in his collection come from a different time and inspiration, [which are] more acceptable to foreigners. The design is based on the concepts that Westerners have of China and closer to Western taste. The elements

> [representative of China] are there: the Chinese tunic, the hats, the application of certain colors like blue and green. You quickly realize that it is China [seen] through the eyes of foreigners.

Jenny hopes to create her own luxury brand, taking into account Chinese sensibilities and aesthetics. Her temporal orientation involves retention and protentions. When Jenny speaks of creating her brand, she approaches an explicit future orientation comprising imaginal anticipations of a determinate future predicated on dissatisfaction with current experience. Her critical evaluation of Western brands' use of Chinese design elements implicitly elevates China's standing above Europe in the future.

Movement from a determinate future orientation to the subjunctive castings of possible futures

Julianna, a homemaker from Hong Kong, recalled her evolving attraction to luxury brand jewels:

> I was crazy about [luxury brand] diamonds, and bought as many as possible … It was all very exciting and my heart pounded furiously every time I touched them. Then I decided to learn about them and took seminars. I felt that I could appreciate my jewellery even more … A Cartier diamond ring [was now] a ring with diamonds of a particular origin, cut, clarity, and carat weight. The brand had a heritage and history, which intensified the pleasure. Earlier, I just liked a particular style or brand. But now I enjoy the details even more. I have the excitement of a collector …

Julianna's initial pleasure derives from the purchase of luxury goods, with diamonds a key symbol of luxury and status. She describes her pleasure as an uncontrollable, positive bodily response to the diamonds and the brands themselves (Russell and Levy 2012). Her subsequent pleasure comes from knowledge; she can now be more discriminating. Over time, Julianna has experienced greater pleasures from the same objects because she has drawn on the flavour beyond flavour. In her descriptions, one recognises both reflection and embodied engagement. Russell and Levy's (2012) observations of how consumers reconstruct memories of an object experienced are similar to Julianna's experience, which in turn informs her dreams of owning more pieces.

Julianna's knowledge of luxury brands is high, but her growing knowledge specifically of jewellery is even higher. She views designer items as superior, and classifies herself as an elitist consumer through her choices. In Throop's (2003: 234) words, we could describe her experience (especially her desire to be a collector of fine diamond jewellery) as exhibiting 'a temporal orientation with a subjunctive casting of possible futures and even possible pasts across the fluid space between a past and a future'.

Present orientation and hyper-responsiveness

Although many of our participants exemplified the concept of savouring, they also experienced the basic pleasures of luxury brands. Others focused primarily on the first-order sensations of owning luxury brands. As our participant Grace said: 'In China, it is every girl's dream to own a luxury brand item that we had seen in fashion magazines. But now it is more than one bag – you need several bags, because all your friends carry several bags.' Grace stresses the now and the unmitigated desire for all things luxurious. In most of the examples below, a present orientation is dominant, although retrospectives glances are made, as well as projective ends imagined. Even in the act of tasting, wealthy and aspirational consumers differ.

Fred, a businessman in Beijing, observed: 'I love all these brands [he was wearing or carrying them during the interview]: Cartier jewellery, Armani suits, and Hermès leather attachés. It is not enough just to be professional, you have to look good as well.' Jim, a businessman in Hong Kong, shares Fred's desire for luxury brands: 'I love wearing all these clothes and jewellery [Cartier rings and diamond cuff links]. I carry Hermès bags – they are elegant. I prefer to go to parties where there are film stars and other celebrities. You have to dress up to fit in with this crowd.' Jim and Fred's display of social status, via the luxury jewellery and clothes they wear, manifests their desire to climb social ladders.

Conclusion

Our participants' responses reflect the depoliticised nature of global brands and their consumption as central to economic development. By using luxury brands, participants signal not only social status, but also their personal status, as individuals with knowledge, modern values, and aspirations. The major dimensions identified in our study are temporality, spatiality and sociality. Although our data focus on luxury consumption experiences, these dimensions are applicable to other experiences as well. In China, savouring focuses on the continuation of the object in the mind, a time in which the significance of the work unfolds (Frijda and Sundararajan 2007), allowing for the ability to sustain memory of the affective experience, in comparison with past experiences. Savouring also allows for emotional detachment (Russell and Levy 2012), necessary for grasping nuances. During this state, thoughts and emotions connect, allowing for self-reflexivity, even as the self is carried away in reverie.

Time is a complex factor with many dimensions. As Throop (2003: 234) suggests, in addition to Heidegger and Husserl's understanding of time in terms of pretension and retention, at least four different temporal orientations may structure the experience of self and world. In our study, the orientation to the present, comprising unfulfilled protentions as open anticipations towards an indeterminate future, is the least encountered. The explicit future orientation, of imaginal anticipations of a determinate future predicated on the residues of past experiences, is the most encountered. Between the two extremes, we identify the retrospective glance that entails the narrative of beginnings, middles and ends over the already elapsed span of a delimited field of experience, in addition to the subjunctive castings of possible futures, and even possible re-imagined pasts, across the fluid space between then and later. We extend Throope's nuanced understanding of time by identifying variations within each category of time.

In conjunction with time, a spatial perspective emerges. The past is primarily viewed by our participants as recent, referencing the Mao period, when China ranked poorly as a force for culture and artistic endeavor, given that class mobility was under siege. Participants referenced the distant past, in ancient and medieval times, when China experienced a golden age of artistic expression and trade. When participants contextualise China in relation to art and culture, regardless of the era, Europe, and in particular France, Italy, and Switzerland, are seen as sources of brands that define a luxurious life; China is relegated to a lesser status. Our findings suggest that luxury experiences are on a continuum. On one end is the completely discursive realm, in which sensations and emotions are translated as a variant of self-talk and connect bodily experiences to symbolic worlds of meaning (Bahl and Milne 2010). As Heidegger (1996) notes, we develop an attitude toward the world according to a future orientation of goals and desires where the narrative is clearly structured. At the other end is an embodied realm, in which the entire focus is on the body and bodily sensations with a present orientation, with few internal conversations and narratives of meaning, even as an individual's past experiences remain influential: what James, Husserl, and Schutz (Throop 2003, 2011) term retentions (looking back) and protentions (anticipating). Between the two endpoints are a number of variants. One can monitor sensations with or without naming them. One can state the name (sensation) and look for causes, or one can build narratives and move entirely into the discursive realm. Imagination mediates between and synthesises sensations and feelings on the one hand, and discursive thought and conscious linguistic and visual categories on the other (Joy and Sherry 2003; see Chapters 14, 15 and 16, this volume). The spatial and temporal dimensions are inextricably entwined, even as they are structured by socio-political contexts (Dong and Tian 2009).

With the acquisition of a luxury brand item, a consumer's social and cultural capital is instantly heightened, offering entry to a rarefied world in which they are shielded from condescension. For the truly wealthy, of course, such social signifiers are irrelevant: their place at the apex of their social strata is secure. As some of our participants revealed, luxury brands are of particular appeal to those who, to varying degrees, are less sure of themselves and their place in the prevailing social structure. Luxury brands provide the knowledge and quality that moneyed consumers from China long for. Bourdieu's (1987) description of how generational socialisation into luxury allows the reproduction of class structures does not truly apply to China. Like Üstüner and Holt (2007), we argue that cultural capital is acquired through education at prestigious schools in the UK and the USA, where children and young adults will be immersed in upper-class social constructs, with cultural capital, as embodied by their possession of luxury brands, serving as their means of entry.

In consumers' second-level savouring responses we see marked variations between Western and Chinese culture. In the past the concept of 'the West' (ranging from European aristocratic traditions to American liberal democracy) has featured prominently in the Chinese imagination as representing modernity and civilisation (Fong 2011). For consumers in Hong Kong and Mainland China, Western luxury goods represent a melding with the world, a blending of cultures in which traditional Asian motifs are subsumed within Western style. Unsurprisingly, China and Hong Kong are experiencing a resurgence of fealty to Asian designers who offer authentic expressions of Asian representations of beauty.

The action component, the embodied actions taken by individuals, is immediate: Embrace the brand, or do not buy it. The action component is downplayed in detachment, although the individual is action-ready, whether mentally or in an embodied sense. Detachment is derived from contemplation, rather than from contempt or indifference. A consumer may be spellbound but can nonetheless hear, touch, smell and see the experience. While there is less arousal in such situations, thanks to pragmatic self-restraint, individuals are receptive to information from their environment, which allows them to make sense of the process, and to prolong their experience. The search for harmony also celebrates differences and diversity in the creation of a new unity. Consideration of cultural nuances is essential for understanding the luxury brand experience within China. Zhou and Belk (2004) advise corporate brand managers: 'When in China, advertise as the Chinese think. We add: "…and feel."' While social status signalling is important in the appropriation of luxury brands, other cultural constructs, such as tasting and savouring and its accompanying temporal and spatial aspects, are equally important for a full understanding of the complex processes at play.

Acknowledgement

The authors would like to acknowledge receipt of a grant from the Social Sciences and Humanities Research Council of Canada, no: 435-2013-1211.

9

FOOD, ALCOHOL AND THE 'IDEAL' HOME IN URBAN CHINA

Chen Liu

Introduction

In his edited book *Home Possessions* (2001b) Daniel Miller highlights the importance of material cultures in shaping domestic social relationships and practices. For Miller 'home' is a dynamic place which is made through socio-material relations, encounters and everyday practices of people and things. Later, Miller (2002) develops his understanding of the social relations between home, interiors and the people as 'accommodating'. 'Accommodating' speaks to a structure of being at home 'in the sense of a place to live'; involving a constant reciprocal process 'in the sense of an appropriation of the home by its inhabitants and of the inhabitants by the home'; and the expression of 'a sense of willing' and 'an agreement to compromise on behalf of the other' (Miller 2002: 115). Home not only means finding or creating a place to live, but also refers to the practice of home-based consumption that involves the objectification of the temporal and spatial interrelationships between the material home, people and things.

In this chapter I engage with Miller's (2001a, 2002) theorisation of home in parallel with Gregson's (2007: 21) argument that dwelling is about 'inhabitation, cohabitation and the practices of habitation ... achieved through an ongoing flow of appropriation and divestment; through acquisition, holding, keeping, storing and indeed ridding'. In this sense, dwelling is a concept in relation to being or feeling at home accomplished through 'living amongst certain things and doing things with and to these things ... [and] what is done with and to ordinary everyday consumer objects' (Gregson 2007: 177). Such arguments emphasise the importance of home-based consumption, and highlight that material practices of (in)appropriation both maintain dwelling (being or feeling at home) and make it geographically more extensive and mobile in the contemporary world. This process of accommodating is not only spatial, but also temporal. People's doings with and to things can make the process of accommodating transient. Nonetheless, through attending to processes of domestic accommodation the identities and social relations of love, care and devotion that sustain the mundane family life within the domestic interior can be presented and narrated.

Employing the concept of 'accommodating', this chapter focuses on the process of home-based consumption of food and food-related things. This chapter aims to build a geography of making 'ideal' homes through an analysis of domestic food (and alcohol) cultures. The questions of how the process of accommodating is accomplished through everyday food practices and how the social relations between home, food-related things and people are shaped by and shaping the accommodating process will be discussed though four food stories from middle-class households in urban Guangzhou. Whilst the importance of everyday food consumption is understood diversely in different social contexts, theories of the everyday have been constructed from a Western perspective. Empirically, everyday experiences – both in the Global North and Global South – have overwhelmingly been researched from a Western perspective. This chapter provides an antidote to such orthodoxies and offers insight into domestic food (and food-related) practice from a Chinese perspective.

For Chinese geographers, urban consumer cultures have generally been equated to restaurant cultures or eating in public spaces (Cai *et al.* 2004, 2006). Indeed, social studies of eating out and family life in Chinese context (e.g. Lozada 2000; Min *et al.* 2004) have suggested that the very notion of family, and family life is extended into and played out in public spaces (see Chapters 2 and 4, this volume). Other research into food and Chinese family life have primarily focused on changing values (Yen *et al.* 2004; Pingali 2007), norms (Martin 2001; Raven *et al.* 2007) and manners (Jing 2000; Jinag *et al.* 2006). Among these studies food is generally considered as a social and cultural force that constructs and reconstructs family roles, rules and rhythms. This work has also rightly indicated that food plays a significant role in making Chinese families, constructing and re-constructing family relationships and negotiating the meanings of home and family for Chinese people.

Despite this progress, studies of the relationship between food and family life in China have overwhelmingly ignored domestic spaces, as key constituents of contemporary urban life as well as material aspects of home and family life and their role within wider, value-laden imaginations. Non-Chinese work on the material spaces of home, and how these construct food practices, can nonetheless usefully be extended into Chinese scholarship. Moreover, although studies have highlighted the importance of the family relationship in the establishment of new urban cultures in China through the lens of food, the questions of how Chinese people endow meaning within their dwelling homes through food practices and how they create their own everyday geography of food within the material home are often taken for granted. This chapter corrects that absence. I present research conducted from an approach that critically prioritises everyday practices, with the purpose of exploring how people use food consumption to establish their own forms of urban life in and from the domestic sphere. The empirical material presents food (and drink) stories from four families to explore people's relationship with food or food-related things: building foodspace, settling, storage and wasting. Tracing the process of home-based food consumption through an ethnographic approach, this research provides a touchstone for understanding emergent developments in Chinese urban life. As indicated by the existing works on Chinese consumer culture, domestic consumption can be considered to be a reflexive and critical narrative that signifies agency and individualisation and an experience of emancipation from the de-commodified past in urban China (Davis 2005; Yu 2014).

The research presented throughout this chapter includes household data gathered through guided home tours, and narrative 'food stories'. The home tours were undertaken using two methods. One was 'exploring' the informants' homes, especially their kitchens and the dining areas. The other method was sharing familial food practices, including going food shopping, and sharing cooking and eating experiences with informants in and beyond their domestic spaces. After each tour, in-depth interviews were undertaken focusing on the meanings and imaginations of home and family, everyday food consumption/activities, food habits, family relations and brief biographies of the family were collected alongside existing photos or videos of family meals or celebrations. To protect the personal information of the informants, all names shown in this research are pseudonyms which are based on the informants' children's childhood names according to the idiom in China – people tend to use the format of 'a child's name add kinship term(s)' to address their relatives (Shen 2013).

The 'ideal' home in urban China

Home is a site of aspiration and a way of self-expression (Clarke 2001). The imaginations of the 'ideal home' are 'not just trivial fantasies about a perceived aesthetic style or associated social aspiration, rather they offer an idealised notion of 'quality of life' and an idealised form of sociality' (Clarke 2001: 28). In practice, home carries 'the burden of the discrepancies between its actual state at a given time and a wide range of aspirational "ideal homes" that are generated out of much wider ideals that a household might have for itself' (Miller 2001b: 7). Yu (2014) suggests that urban consumer culture in China has changed from a monotonous and state-planned model in the 1990s into a globalised and marketised formation in recent years. In other words, urban China has been transformed from a socialist society into a consumerist one. According to Elfick's (2011) research, contemporary Chinese middle-class consumers invest a lot of time and money on the decoration of their homes as a way to express their own aesthetics and understanding of the meanings of home (see Chapters 4 and 8, this volume). The preference of the global, modern, stylish and personalised home design, which can express their tastes and personal aesthetics, has become the dominant home culture in urban China (Elfick 2011; Yu 2014). Although commercial houses with fine decor have become popular among urban middle-class consumers in Guangzhou, many of them are still keen on redesigning and/or redecorating their homes in order to express their individual tastes (*Guangzhou Daily* 2009).

Such personal aesthetics are often intertwined with people's understandings of modernity with the link to a global/international way of life. For example, modern home cultures were embraced by Xiao Mi's parents through the medium of IKEA home design and furnishings, which they considered modern, stylish and pragmatic with reasonable prices:

Xiao Mi's father: the products in IKEA fit our styles. The IKEA style is modern and simple which fit young people's tastes and aesthetics. And the price is affordable for us. I like the stylish, comfortable and cheap things for IKEA … our generation likes adopting international ways of life which make our life convenient.

Xiao Mi's mother: and their design is pragmatic. You see, this table is easy to assemble and useful

(quote from a recorded interview with Xiao Mi's parents)

Apart from expressing cultural taste and identity at home, people also used the word 'comfort' (*shushi*, or *shufu* in Chinese) to describe their ideal homes. This extends the emphasis placed on practicality by Xiao Mi's parents in the quote above into a broader and more complex affective response. Comfort was important to feeling at home. Based on her ten years of fieldwork in Shanghai, Davis (2005) indicates that the desire for a materially comfortable home and the pleasure of home renovation are central to their ideals of home design. In her opinion, the process of making home can be considered as a reflexive and critical narrative that signifies agency and individuation and an experience of emancipation (from the de-commodified past) in urban China. For some participants of this study, achieving comfort requires that modern domestic spaces and furnishings engage with traditional Chinese views on domestic space. For example, Xiao Mi's father combines the traditional idea of *feng shui* (Chinese geomancy) into his redesign of his new home in order to make it comfortable (I will discuss further details in the 'making the dining area' section). More generally, the idea of 'comfort', according to the households that I researched, often referred to the colour, shape, size and 'textures' of the home and its interiors. For example, Xiao Tian's parents link their longing for a comfortable home to both the furnishings and home technologies:

Xiao Tian's mother: [our consideration of home design is] how to make our home comfortable. Hence, we didn't buy a wooden or leather sofa, instead, we bought a fabric sofa, as the texture of fabric is softer and makes us feel comfortable. We also bought the lights with softer lights in both dining room and living room. Also, all of the furnishings and paints are environment friendly, because we want a safe living environment …

Xiao Tian's father: and because we planned to have a baby when we purchased this flat, we designed our home with more colours and painted some cartoon figures on the wall. These make our home warmer
(quoted from a recorded interview with Xiao Tian's parents)

That is, for Xiao Tian's parents, comfort is at the core of their home design and their consumption of home interiors. This idea of comfort is a means to accommodate their imagination of an ideal home in their real home (see Chapter 4, this volume).

The following sections now turn to discuss the ways in which these ideas of the ideal home are achieved through the process of accommodating food and food-related things within the real and actual homes. The empirical evidence illuminates how people place domestic food cultures at the centre of their accommodating and how they connect their material home and family relationships emotionally through food consumption.

(Re)designing the cooking space

Everyday food work in the kitchen is 'organised by, through, and around a physical landscape of material possibilities' (Shove *et al.* 2007: 37). Kitchen design is concerned with those material possibilities. In order to explore how people design their ideal kitchens in contemporary Guangzhou, I use an account of 'open' kitchen design in Xi Xi's home, illustrating how the kitchen design intersects with the social, cultural and economic domestic life of Xi Xi's family.

Xi Xi's family moved in to their new home before the Chinese New Year's Day in 2014. Because this new home was a second-hand flat, Xi Xi's parents refurbished the entire flat once they decided to move in, as they wanted to make it their own home with their design. The first and most important consideration during the process of their home-based consumption is the finances of the family. Although Xi Xi's parents crave a completely new home with new furnishings, equipment and other things, they cannot realise their dream because of their financial condition:

> My first consideration is 'does the kitchen need to be refurbished'. You know, the cupboards are new. So, we only replaced the old appliances. Secondly, I have considered our financial condition at that time. You know, we bought this apartment with a mortgage. Refurbishment needs a lot of money as well. If we spend too much on the apartment, how can we go annual travelling? And how can we afford for our daughter's education? … And finally, I have thought about the question of 'do we really need a completely new kitchen'. So, we didn't build a new kitchen, instead, we only renewed it … we don't plan to upgrade it in these five years, but maybe (do this) later, if we have enough money. And this depends on the kitchen itself. If we kept it well, we won't upgrade it
>
> *(quote from a recorded interview with Xi Xi's father)*

With the limited budget, Xi Xi's parents accommodated their understanding of modern domestic life into their new kitchen, as far as possible. For Xi Xi's parents, the design of open kitchen represents their aspiration of a new modern home (see Figure 9.1):

Xi Xi's mother: we want to add some modern elements in our apartment. As this flat is second-hand, before we moved in, we had redesigned the rooms. You see, we bought the new dining table and chairs, and repainted the wall. I think these new things and colours may make my home more modern.

Chen: What do you mean by 'modern'?

Xi Xi's mother: I think it means stylish and up-to-date. I think a modern home should have stylish designs and high-tech appliances.

Xi Xi's father: I think our open kitchen is a modern design. Originally, the kitchen was closed, but the previous owner of this flat broke one of the walls around the kitchen to make the space of the flat look bigger.

Xi Xi's mother: This design is smart! I never feel lonely when I cook. I can watch TV in the kitchen, and talk to my husband when I am in the

FIGURE 9.1 The open kitchen in Xi Xi's home and the double-tap sink in the kitchen

Source: Chen Liu

> kitchen, as it is open! The only problem is that the open kitchen may make the cooking smell diffuse everywhere in our flat. So to maintain the open kitchen, we replaced the old kitchen ventilator with a more effective one
>
> (quote from recorded interviews with Xi Xi's parents)

It is evident in this quotation that for Xi Xi's parents keeping and upgrading the open kitchen not only expresses their understanding and aspiration of modernity, but also helps to transform the boring and stressful work of cooking into more joyful family activities. More specifically, Xi Xi's parents tend to connect their understanding of the modern home with new technologies. Technology plays an important role as a liberating force in the experience of contemporary consumption (Miles 1998: 70 – 72). The everyday practices of and with such food technologies are important ways of making and changing the modern home (Shove 2003; Shove *et al.* 2007). In Xi Xi's home, both ventilation and cooking technology were adapted to make their open kitchen function better: 'We installed a high-power ventilator in the kitchen to reduce the cooking smell, and upgraded the gas stove, in order to reduce the smoke' (quote from a recorded interview with Xi Xi's father).

Two-way relations between the family and domestic food technologies are apparent here. On the one hand, Xi Xi's parents sought to achieve their aspiration of an 'ideal home' through designing their open kitchen. On the other hand, in fact they then become 'slaves' to this design, working to accommodate it in their home; they moved things around and worked out solutions to reduce the cooking smell, in order to achieve their ideal of a modern kitchen design. Through their design and redesign, the kitchen in Xi Xi's home becomes temporally and spatially dynamic. These dynamics extend beyond the outline of 'having-doing' dynamics emphasised by Shove *et al.* (2007), which highlight the precarious relations between kitchen possessions and kitchen practices. The dynamics in this kitchen involve complex relations between the material possessions in the kitchen, family practices, its owners' imagination of what a kitchen should be, and their finances (purchasing power). It is these dynamics that the open kitchen, with new ventilation and gas cooker, objectifies.

Making the dining area

The dining area is always recognised as more than an important place for family meals, and instead as a multifunctional domestic space that enables cooking, eating and living in the same place. In her overview of kitchen in the Western world, Johnson (2006a) indicates that the small 'kitchen-meals area' has expanded into a 'kitchen-meals-family-games precinct' where 'most family and entertainment activities' take place (Johnson 2006a: 129). Beyond Western societies, the kitchen is also often a multifunctional space where family activities occur and where household social norms and orders are performed. For example, Buckley (1996) points out that the kitchen in Japan is more than a cooking space or a site for completing family meals, women's domestic work and children's education; it is also the representation of the state's call to combine kitchen-dining-living areas and a construction made under the predominantly masculinist principles of gender, family and work.

Drawing on the ethnographic observations in Xiao Mi's family home, this section underpins the importance of the multifunctional dining area in home design and domestic food practices. Xiao Mi's family moved to their new flat – a newly built flat in a high-rise residential building located in a gated community – in October 2012. According to Xiao Mi's parents, all the flats in their community were built and furnished according to national standards. After they purchased this flat, Xiao Mi's parents transformed the separate kitchen, dining room and living room into a more open and fluid cooking-dining-living area. According to Xiao Mi's father,

> When we bought this flat, the structure is not like this. There was a wall here, separating the kitchen from the dining area, but we removed the wall, and replaced a door … that design had bad *feng shui* which would do harm to my family. We changed the structure: sealing up the old door, and opened a new door here. Therefore, the geomantic omen is better, and it makes a shorter way to enter in the kitchen as well
>
> *(quote from a recorded interview with Xiao Mi's parents)*

Thus, whilst the national standards propose separating the dining area and kitchen, the owners – Xiao Mi's parents – challenge this official idea of home design through their own creative homemaking. They did this based on their own aesthetics, knowledge of home, as well as their own sense of daily food practice routines. In the process of fashioning a kitchen-dining-living area, they make their home more homely. After changing the house structure, Xiao Mi's parents placed the dining table, tablecloth and chairs within their open dining area. The tablecloth (see Figure 9.2) was bought by Xiao Mi's mother during her holiday trip:

Xiao Mi's mother: I liked the ethnic style several years ago. I bought many tie-dye cloths when I travelled to Feng Huang (a Miao[1] county in Hunan province). I didn't know what these cloths could be used for, until I found we need a tablecloth for the dining table. But now I don't like the texture of this kind of cloth and I am not interested in the ethnic style any longer … I prefer a minimalist lifestyle.

Chen: Why not replace this tablecloth and other things with minimalist ones?

Xiao Mi's mother: I cannot focus too much on the home design. I should care for my daughter. I don't have enough time to think about it. And,

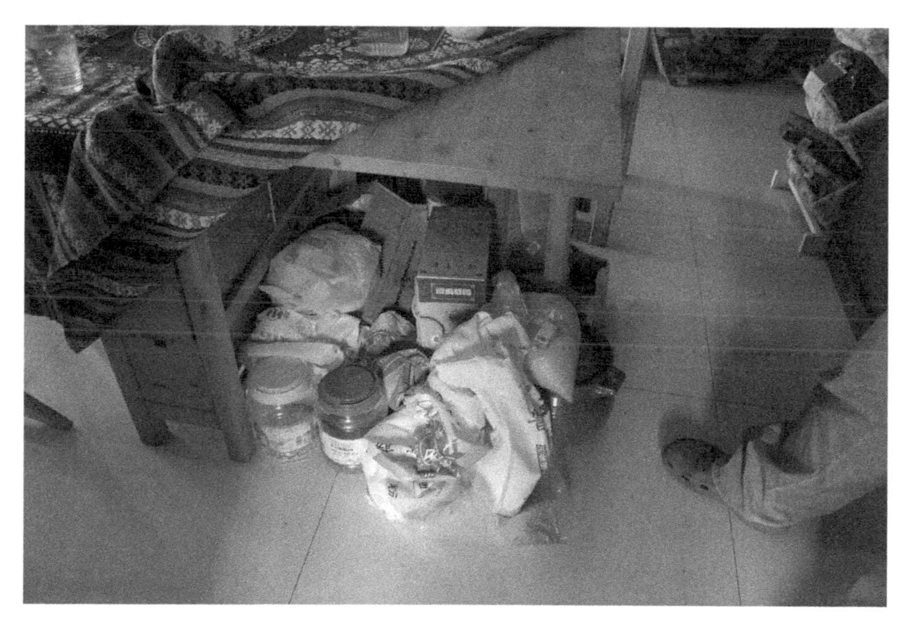

FIGURES 9.2 and 9.3 The dining area in Xiao Mi's home

Source: Chen Liu

> when my parent-in-laws come to live with us, this home is no longer my home, but becomes our home. I don't want to change it only for expressing my aesthetics
>
> (quoted from a recorded interview with Xiao Mi's parents)

This quotation suggests that the dining area, for Xiao Mi's mother, is more than a place for eating, but rather, a place to display her cultural tastes and her changing understanding of home. In this case, her aesthetic expression, and its changes over time, is somewhat subordinated to the wider family's needs and desires. Tellingly, the tablecloth that decorates the dining table does not simply express her personal taste but filters this through other domestic dynamics, in particular intergenerational relations. The 'consumption' undertaken through this space is not that of an individual consumer but a more complicated imbrication of personal biography, family/household relations and wider cultural discourses and developments.

The dining table also serves as a perfect place for food storage (see Figure 9.3). During the home tour, Xiao Mi's mother lifted up the tablecloth and showed me the food – including several bags of flour and rice, two large pots of oil and some boxes of dried vegetables – stored there by Xiao Mi's grandparents:

Xiao Mi's mother: My parent-in-laws like to eat the food from their hometown. So, they always bring a lot of hometown food to our home. But our kitchen is too small. Therefore, they put them under our dining table.

Xiao Mi's grandfather: We cannot find such things in Guangzhou … for example, the linseed oil.

Chen: I think you can find this kind of oil in the supermarket.

Xiao Mi's grandfather: But it is too expensive! [It is] 500 ml for more than 100 CNY! But in our hometown, 100 CNY can afford 2 kilograms. I brought two plastic pots of oil … [and] put them under the [dining] table

(quote from the field diary)

Therefore, this dining area is made multifunctional by both Xiao Mi's parents and grandparents, in order to maintain the different food preferences and tastes of these two generations.

Settling culinary cultures in the bedroom

This section focuses on the food stories of Xiao Tian's parents, in order to explore people's practices with/to food-related things in 'inappropriate' places. Xiao Tian's family lives in a flat in a newly built gated community. Different foods and food-related things are deliberately placed in different rooms in Xiao Tian's home. Xiao Tian's mother put her teapots, cups and tea in the cupboard between the living room and the dining area, because she loves to drink tea and likes to buy tea-related things. In contrast to Xiao Tian's mother hiding her collections in the cupboard, Xiao Tian's father displayed his collections of wine and Chinese liquor at the entrance of their home. According to Xiao Tian's father, these bottles of alcohol on the one hand can be home decorations, and on the other hand reflect his taste:

> I like to collect alcohol. I display my collections on the shelf, although some of them are cheap and common. They are about my taste of life and my hobby. And, the bottles can make the entryway look better. When my friends or my neighbours come, we can talk about my collections and try my collections
>
> *(quote from a recorded interview with Xiao Tian's father)*

It is evident that the accommodation of the alcohol collections at the flat entrance reflects their role in relating to absent others – the friends of Xiao Tian's father. These bottles become a decorative materialisation of the hospitality offered to his friends in the home. However, the most treasured alcohol is not put on the wine shelf at the entrance; rather, it is accommodated in Xiao Tian's parents' bedroom. After a brief introduction to wine and liquor knowledge, Xiao Tian's father asked me to come in his bedroom to see a box of Mao-tai (a famous kind of Chinese liquor, see Figure 9.4), which he sees as perhaps his most important culinary possession:

> This box of Mao-tai is my treasure. Because this is for my daughter, I put it in my bedroom. When I was on a business trip to Guizhou, I bought this box of Mao-tai. I want to give it to my daughter as her wedding gift … Our wedding ceremony and the celebration of my daughter's 100 days after birth was in the same day. In the ceremony, my wife and I promised to keep this box of Mao-tai until my daughter's wedding ceremony. In order to guarantee the quality of Mao-tai, I bought it from the Mao-tai distillery in Guizhou province. Each bottle of Mao-tai costs me 2,000 CNY. It is expensive, right? But it is worthy! I think this box should stay in my bedroom for more than 20 years.
>
> *(quoted from a recorded interview with Xiao Tian's father)*

Xiao Tian's father 'hides' this box of Mao-tai in his bedroom, in order to keep his love and care towards his daughter in a very private space. The segregated accommodation of alcohol in Xiao Tian's home constructs a boundary between the family and non-family spheres: for Xiao Tian's father, the alcohol displayed on the shelf is for presentation in front of his friends; while the Mao-tai stored in his bedroom is only for his daughter and her future life.

FIGURE 9.4 A box of Mao-tai with the production date in Xiao Tian's father's bedroom

Source: Chen Liu

Disposing of food waste

Food waste is associated with food production, distribution and consumption practices, and is crucial to understanding our living world and more-than-human matters (Alexander *et al.* 2013; Evans 2012; Watson, 2013). In domestic spaces, food waste is not simply a discard that ends its life in the bins, but also an engagement with the broader practices of household food provisioning and consumption. Considering food waste as a result of food over-provisioning during conventional family meals, and as part of the everyday scheduling of domestic food work in the UK, Evans (2012) argues that food waste is both 'a consequence of the domestic practices' which are 'socially and materially organised' and the negotiation of the 'contingencies of everyday life' (Evans 2012: 42). Watson and Meah (2013) also focus on the relationship between the transformation of food into waste and domestic food practices in UK society. They argue that the policy issue of reducing food waste needs to be rooted in the organisation of the quotidian routines of food practices, and suggest that interventions to reduce food waste can be enhanced by appreciating how food becomes waste through everyday practices. In summary, such studies understand food waste as an important part of the routinised food provisioning and consumption in domestic spaces which are in turn socially, politically and materially organised. Focusing on the divestment and disposal of kitchen waste, this section examines how household food waste becomes the fallout from the everyday doing of family life and how it is engaged with complex networks that include people, food waste, non-human entities, public policies and the space of home, based on the food story of Qi's home.

In order to reduce waste disposal costs and to promote a green lifestyle, the government of Guangzhou enacted the *Interim Provision of Guangzhou Municipality on the Administration of Urban Household Waste Classification* in 2011. Since then, the Urban Management Office of Guangzhou began to hand out kitchen waste bins to every household in Guangzhou and set up sortable garbage bins in each community. However, many of the residents in Guangzhou neither received the kitchen waste bins nor could find the sortable bins in their communities. Therefore, instead of disposing household waste according to local policy, many people create their own procedures for food waste, guided by their own concerns about thrift and the proper ways to divest themselves of inedible food. Qi's family moved into their current home at the end of 2012. Although Qi's family did not receive the standardised kitchen bin allocated by the local council, people in this family consciously undertake garbage classification. In Qi's home, there are two waste bins: a wastepaper basket in the study room and a big trash bin located inside the kitchen for collecting kitchen waste. Also, Qi's family members put old newspapers, cartons, cans, glass bottles and other recyclable waste on the balcony, waiting for the regular collection by the 'binman'.[2]

Apart from the everyday reduction and classification of waste, Qi's father developed a unique way to dispose of the household food waste: composting it. When I asked Qi's father to show me his special or important things related to food, he pointed to a plant on the balcony (see Figure 9.5):

> These plants are about my creative use of food waste and reveal my green attitude towards my life ... On the bottom of the plant, you can see some peasecod, the skins of fruits and other inedible parts of vegetables. When I put these food wastes in the pots, they become fertilisers. They not only fertilise the plants, but also reduce the kitchen waste! These food wastes become useful things
>
> *(quote from a recorded interview with Qi's family)*

However, this kind of food waste disposal sometimes causes family disputes: over the disgusting smell that some waste has as it decomposes and fertilises the plants; and over other hygiene problems resulting from the usage of food waste as fertilisers:

Qi's father: I think this recycling is good for our environment and it saves money. You know, a huge amount of money was spent to deal with food wastes. However, other people in this family are opposite to my idea! They think this rubbish is unhygienic and disgusting. Anyway, I keep on doing this, because it does more good than harm. Isn't it?

Qi's mother: But it still smells bad and attracts many flies!

Qi's father: Not so bad.

FIGURE 9.5 A plant that is fertilised by food waste in Qi's home

Source: Chen Liu

Qi's mother: Because you never stay around your plants. The only thing you do is put the kitchen waste into your flowerpots.

Qi's father: I put them on the balcony. No one would stay around them all the time! Come on! This is a green idea of food waste disposal

(quote from a recorded interview with Qi's family)

Therefore, for Qi's family, food waste disposal not only reflects people's understanding of the relations between food and our living world, but also plays an important role in the doings of family life and family relations. Furthermore, this creative way of dealing with food waste is both a response to the inefficient implementation of the policy on garbage classification in Guangzhou, and also a (contested) performance of people's moral considerations on wider environment and ethical issues.

Conclusion

This chapter portrays the urban life of middle-class people in Guangzhou through the lens of daily food consumption. It helps to understand how people practise their own understandings of urbanism, modernity and 'proper' family relationships through everyday practices of/with food or food-related things in the domestic sphere. The main argument of this chapter is that people's everyday food practices are inscribed in and motivated by material processes of accommodation and homely dwelling.

Food (and alcohol) are highlighted as key aspects of domestic life, implicated in both imaginations of ideal homes and in the practices that perform and reshape these ideals in the material making of real homes. For the respondents, ideals and practices of home often seek to articulate desires for modernity with desires for comfort. These desires play out through the food-related material fabrics of the home, ranging from the design of the domestic space to the installation and use of food-related appliances, furnishings and decorations, storage and disposal of food. People in contemporary Guangzhou accommodate their culinary cultures into their actual domestic spaces to achieve their aspirations of an ideal home and as a form of self-expression: Xi Xi's parents design and redesign their open kitchen to accomplish their modern ways of life; Xiao Mi's mother uses the Miao style tablecloth to accommodate her preference of ethnic cultures; Xiao Tian's father uses his alcohol collections to convey his love and his relations with others; and Qi's father composts food waste in flowerpots in order to sustain his green lifestyle. However, the process of accommodation is not only about how people use things, but also about how the physical home and its interiors do things to people: the design of the open kitchen necessitates the cost of a high-powered ventilation system that increases the consumption of electricity; and Qi's father composts food waste in flowerpots because it is impossible to locate gardens within home spaces in their high-rise residential buildings. Also, these processes of homemaking are strongly linked to compromises with visible and invisible others. Together, these cases have confirmed the three central arguments about domestic 'accommodation' identified by Miller (2002): home is a place for everyday living; there is a reciprocal relationship between the material home and its inhabitants; and the importance of visible and invisible others in this process of material homemaking.

This chapter also advances Miller's (2001a,2002) notion of accommodation through exploring the strong link between food cultures and home spaces in Guangzhou's socio-cultural context. First, accommodating, as an everyday practice, is both spatial and temporal. Because we want to use things to narrate our ongoing life and identities and because the relations between what we have and what we can do are precarious (Shove *et al.* 2007), the process of homemaking never comes to an end. As shown by the story from Xiao Mi's home, Xiao Mi's mother keeps on changing the home style to better represent her changing aesthetics and identities. Secondly, the juxtapositions of different ways of life are displayed through the material processes of accommodating. As shown through the story of Xiao Mi's family, the grandparents' preference of food coexists with the mother's cultural tastes in the dining area through their different usages of the space. Accommodating, thus, is more than an assemblage of social relationship, material home and things, but rather a process of negotiation and cohabitation of people and things. And finally, the practice of accommodating is not only materially, emotionally and socially organised, but also governed by public policies and influenced by wider society. The researched homes in this chapter are not only accomplished and/or obstructed by the relations between home, things and people, but also standardised by national or local policies and influenced by the commercial trends of the housing market in

Guangzhou. Therefore, accommodating is an articulation of the material and the practical which shapes and is shaped by the social, cultural, emotional and political relations between people and home interiors, homespaces and the city.

My findings also suggest that domestic food practices are not constrained to the kitchen and dining area. Home-based food consumption is not only about cooking and eating, but also about other doings with and to food. Undoubtedly, the kitchen is the most important food space within the home, as the major parts of domestic food work, including cooking, preparing, storing and cleaning up, take place in the kitchen. Due to these practices, the kitchen is both a container for household food work and a site that designs everyday domestic rituals and social relations. However, the material cultures of food in the domestic sphere can also be located in every corner of the home. In short, this exploration of accommodating food cultures has illustrated that the space of home becomes multifunctional and fluid/porous in contemporary Guangzhou.

Furthermore, the accounts presented in this chapter highlighting processes of domestic accommodation of culinary cultures contribute to the understanding of the meanings of modernity in urban China. The term 'modern' in the ideal modern home is not only about the bridge between Western and Chinese cultures (Cui 2012), but also characterised as the combination of contemporary living with the 'comfort' of caring for others (both of the family members and the invisible others) and oneself. This modernity is forged under the standardisation of the state council and the influences of commercial trends. However in practice, people's imaginations of the 'ideal home' are carried out or abandoned in their real homes due to the influence of their finances, technologies, house structures and social relations with others who co-reside with them or who are potential visitors.

Notes

1 Miao is a minority ethnic group in China
2 In China, the 'binman' is a person who makes money from buying and selling recyclable waste. The 'binman' buys waste paper, cans, plastic bottles etc. from local communities then sells this waste to the garbage station in order to make a profit.

10

POP-UP URBANISM

Selling Old Beijing to the creative class

Amy Yueming Zhang

Introduction

Urban regeneration strategies engaged with artistic and cultural elements are identified as common practices in many Western cities. Artistic and cultural uses are valued in such cases mainly for two sets of reasons: for being part of the knowledge and cultural economy that is increasingly pursued by cities since the early 2000s (Florida 2002a, 2002b; Hutton 2004; Mommaas 2004); and for their ability of changing the perceptions of a city as well as revalorising real estate through directly reusing and aestheticising abandoned properties (Catungal *et al.* 2009; Mathews 2008; Stern and Seifert 2010). This function of art and culture in changing the image of a city or neighbourhood and in revalorising urban land and property values is well documented by many scholarly works, through cases such as art-led gentrification and building cultural flagships for urban regeneration, both before and after the spread of the 'creative city' thesis (Balibrea 2001; Cameron and Coaffee 2005; Evans 2003; Gomez 1998; Ley 2003; Plaza 2008; Zukin 1989).

Despite attention paid to the globally circulating 'creative class' and 'creative city' discourses, the relations between these discourses and cities' adoptions of art- and culture-led urban regeneration strategies, and various effects of such strategies (Atkinson and Easthope 2009; Catungal *et al.* 2009; Mathews 2014; Peck 2005; Prince 2013), research is still scarce on whether, how, and to what extent art/culture-led urban regeneration strategies are adopted and function in cities beyond the Global North (Harris 2005). Moreover, existing literatures point out that both the practices of adopting and adapting the creative city model, and the actual effects of engaging with art and culture for urban regeneration are heavily dependent on local and national socio-political contexts (Andres and Golubchikov 2016; Boren and Young 2012; Da Costa 2015; Grodach 2012; Harris 2005). This study examines how strategies of using artistic and cultural elements for urban regeneration are carried out 'beyond the West' – in Beijing, China – and discusses the implications of such strategies to this particular context.

The case examined in this chapter is an urban regeneration project conducted in one of Beijing's inner-city historic conservation areas – Dashilar. In this case, district government employs culture-led urban regeneration strategy by incorporating the area into a municipal level and city-wide event: Beijing Design Week. More specifically, a pop-up art and design landscape is created in this area every autumn during Beijing Design Week, with the goal of establishing sustained interests of the creative class – both the producers and consumers of 'creative products'. The ultimate goal, though, is to capitalise on the presence of 'creative uses' to attract tenants who can provide higher rent. In doing so, I contribute to ongoing debates regarding the role of art and culture in urban regeneration, as well as exploring classed dimension of urban politics in China.

Art and culture in urban landscapes

Art and culture have long been incorporated into urban regeneration to '[donate] a saleable neighborhood "personality"' (Smith 1996: 19) to run-down areas where real estate investors aim at attracting middle-class consumers to (Hackworth and Smith 2001; Smith 1996; Zukin 1989). Such a function of artistic and cultural elements is argued to be the result of two combined forces that have their roots in the case of Soho, New York City, in the 1960s – 1970s (Shkuda 2015; Zukin 1989). On one hand, Soho artists, in their efforts of legalising residential or mixed use of loft buildings, advocated the idea that the atmosphere created by artists and their living-working milieus could attract middle class consumers and hence drive real estate development (Shkuda 2015). And on the other hand, the popularity of 'loft lifestyle', which validated Soho artists' claim, was based on an 'aesthetic conjuncture' (Zukin 1989: 15) occurring in the 1970s, when a societal valorisation of art coincided with domestic appreciation of industrial design and aesthetic (Zukin 1989). The case of Soho provides a strategy to large real estate investors, namely relying on the art community to 'smooth the flow of capital into neighbourhood' (Hackworth and Smith 2001: 467). Through such 'artistic mode of production' (Zukin 1989: 176), real estate investors not only found a way (gentrification) to replace the 'slash-and-burn' mode of urban redevelopment that is often strongly opposed by the public, but also to easily change the image of run-down neighbourhoods and repackage those areas as commodities catering to middle-class consumers (Deutsche and Ryan 1984; Podmore 1998; Smith 1996; Zukin 1989).

The relationship between elements of art and culture and urban regeneration have evolved to a new stage since the early 2000s, with the introduction of the 'creative class' thesis (Florida 2002a, 2002b). In Florida's argument (2002a, 2002b), artists and people who work in cultural sectors are part of a 'creative class' that cities need to attract in order to thrive in post-industrial economic transitions. At the same time, artistic and cultural elements are identified as essential to the type of environment that is attractive to the creative class: 'Places are also valued for authenticity and uniqueness. Authenticity comes from several aspects of a community – historic buildings, established neighbourhoods, a unique music scene, or specific cultural attributes' (Florida 2002c). Thus, to city governments who make policies in accordance with the 'creative class' and 'creative city' theses, having urban landscape with a saleable aesthetic is considered as crucial for the city's economic development in a post-Fordist world. Following the script provided by Florida (2002b), art/culture-led gentrification and urban conservation for regeneration become key ways of creating the 'authentic and unique' urban neighbourhoods that are said to be able to anchor the highly mobile creative class talent, and become parts of the policies and strategies issued by city officials (Cameron and Coaffee 2005; Catungal *et al.* 2009; Krivý 2011; Lloyd 2002; Peck 2005, 2007; Rothenberg and Lang 2015).

Creative class/city theses and policies have received much criticism for their neoliberal nature, prioritising investments for luring the creative class at the expense of addressing the need of low-income residents, causing displacement, commodifying art and culture, being tokenistic when engaging with art community merely for the image, and so on (Atkinson and Easthope 2009; Bell and Jayne 2001; Catungal *et al.* 2009; Evans 2003; Mathews 2014; McLean 2014; Miles 2005; Peck 2005). Assumptions underpinning such theses and policies have also been scrutinised by scholars. On one hand, the extent to which the creative class generates economic development is questioned (Markusen 2006; Peck 2005). And on the other hand, the idea that people of the creative class prefer 'authentic and unique' urban built environments is at odds with the fact that Florida also provides a list of recommended components to include in creating such 'authenticity and uniqueness' (Florida 2002b, 2002c). If mass production of the aesthetic lauded by Florida is possible, cities following 'a creative city script' can no longer claim uniqueness (Harvey 2002). By following the creative script, cities not only throw themselves into a zero-sum game, but also run the risk of investing in elements that could suddenly be deemed as 'out of fashion' and undesirable by the creative class as a result of these elements' omnipresent

> increased public subsidies for the arts, street-level spectacles, and improved urban façades, with expected 'returns' in the form of gentrification and tourist income, [that run] the self-evident risk that such faux-funky attractions might lapse into their own kind of 'generica'. The creatives' restless search for authentic experiences may, of course, lead them to spurn such places
>
> *(Peck 2005: 749).*

Such inherent flaws in the consumption-driven logic of the idea of the creative city leads to city governments' constant search for components that would be edgy and cool enough to be consumed by the creative class (see Chapters 3 and 8, this volume). One way of fulfilling such relentless search for novelty is mainstreaming 'underground' and illegitimate art and subcultures piece by piece, such as the selective incorporation of street art and graffiti into creative industries and creative city policies (Colomb 2012; Dickens 2010; Halsey and Pederick 2010; McAuliffe 2012).

More recently, temporary and pop-up uses of vacant urban spaces as artistic or cultural sites, which were previously often regarded as 'marginal', are enlisted in creative city strategies for creating saleable urban landscape, attracting creative class consumers, and eventually achieving urban regeneration and revalorisation (Andres 2013; Colomb 2012; Ferreri 2015; Harris 2015; O'Callaghan and Lawton 2015; Schaller and Guinand 2017). Temporary and pop-up uses function well for the purposes of creative city policies not only because their prior illegitimate status grants such uses the edgy novelty appealing to the creative class consumers, but also due to their ability to induce spatial-temporal imagination and alter perceptions of places by providing materialised potential 'look and feel' to vacant urban spaces (Andres 2013; Colomb 2012; Harris 2015). Through temporary and pop-up uses, presently run-down, vacant, or 'wasted' areas can be imagined 'as playgrounds or workspaces for "creative" entrepreneurs; as milieux that can attract other creative workers and consumers ... or as tourist attractions' (Colomb 2012: 138) in the future, and thus may gain the interests of investors and developers (Andres 2013; Colomb 2012; Harris 2015; Schaller and Guinand 2017).

Existing research shows that there are two slightly different ways of engaging with art and culture for urban regeneration: as token and as catalyst. Arts and culture tend to be treated as tokens when they are perceived as necessary components of an urban landscape for it to be attractive to a certain group of people, which manifests most clearly in the creative class and creative city theses and policies. When being treated as tokens, there may be less of a threat to displace artistic and cultural uses, although they may not necessarily receive the support they need either (Mathews 2014). Arts and culture function more as a catalyst in urban regeneration when their role is to change the perception of specific urban areas and neighbourhoods – to '[tame] the neighborhood' (Smith 1996: 19). In this case, artistic and cultural uses usually face displacement once their service is not in need, as can be observed in art-led gentrification and temporary/pop-up artistic uses. These two ways of engaging with art and culture in urban regeneration are certainly not exclusive and may be combined, depending on different policies made in specific contexts. Also, as many studies point out, the function of art and culture for urban regeneration does not automatically apply to all cases. Such functions rely on particular socio-political contexts, thus local contexts need to be examined when discussing the relationship between art and urban regeneration (Andres and Golubchikov 2016; Grodach *et al.* 2016; Harris 2005). My research thus contributes to ongoing conversations in geographical and urban studies literature on how artistic and cultural elements are incorporated into strategies for urban regeneration and their effects (see Chapter 3, this volume). In the remainder of this chapter, I examine a case where pop-up is used to create a saleable landscape in old inner-city Beijing.

Creating a saleable landscape in Old Beijing

Urban conservation and regeneration in old Beijing

Dashilar is an area located in the centre of Beijing, south of the Forbidden City, Tiananmen Square and near Qianmen. The area is famous for its main commercial street, which was built in the Ming Dynasty and now serves as a major tourist attraction. *Hutongs* (small alleys) in the area are mostly in a mix of residential and commercial uses (see Chapter 7, this volume). Like many other old neighbourhoods in the inner city of Beijing, Dashilar is densely populated with most of its residents living in low-rise, historic courtyard houses. Many of these courtyard houses face the problems of overcrowding, poor maintenance, and unauthorised additions to the original buildings, and need to be renovated and/or restored. As one of the twenty-five historic areas that were designated as urban conservation areas by the municipal government of Beijing in 1999, Dashilar is targeted for reducing population and restoring historic features of its built environment (Beijing Municipal Government 1999, 2002). The municipal government requires renovation and restoration in urban conservation areas to be carried out in the 'microcirculation' (*weixunhuan*) approach, namely on a house-by-house basis, in small and incremental steps, and depending on residents' will (Beijing Municipal Government 2002; Shin 2010).

For example, Nanluoguxiang street – a conservation area in the northern part of the inner city – is one well-known case that adopts the approach of 'microcirculation' for regeneration. As is documented in detail by Shin (2010), in this approach, all residents of one courtyard need to collectively apply for and agree on a relocation plan, which usually results in permanent displacement of the residents and release of dilapidated courtyard houses on the real estate market (Shin 2010: S49 – S50). Most of the emptied and renovated courtyard houses facing the main central lane of Nanluoguxiang are turned into commercial premises catering to middle-class consumers, namely cafés, restaurants, trendy shops and souvenir stores. With a saleable landscape, the street becomes a popular place among both middle-class residents of Beijing and tourists. The result, however, is that the area falls victim to its popularity: people criticise it for being too commercialised, touristy, and overcrowded, especially during peak tourism seasons, which also affects local residents disproportionately in a negative way (Shin 2010). Increasing popularity also turns away some potential consumers of the place, for it has lost the edgy feeling and now acquires a 'look and feel' that is similar to other tourist attractions in historic areas of Beijing. The uniqueness and distinction that gave Nanluoguxiang a monopolistic edge in the market faded away over time (see Chapter 16, this volume).

Being very aware of the case of Nanluoguxiang Street, people who are in charge of regeneration in the Dashilar area claim that they will not replicate what happened to Nanluoguxiang in this area (interview with government representative). The regeneration of the Dashilar area is carried out by two subsidiary companies of a state-owned real estate company (Beijing Guangan Holding Co. Ltd.) that is affiliated with the government of Xicheng District, where Dashilar is located (Dashilar 2013a). Unlike the case of Nanluoguxiang, where a clear plan was missing when courtyard houses were released on the market, the regeneration of Dashilar area is based on a vision of gradually turning the area into a 'creative community' through attracting art, design, and cultural industries (Dashilar 2013a). In order to achieve this goal, the district government managed to make Dashilar one of the sites for hosting Beijing Design Week.

Beijing Design Week is a week-long event featuring exhibitions, design award ceremonies, forums, and design fairs. It usually takes place in the fall and overlaps with the National Day 'golden week' for a few days, and occurs at several sites across the city, including the well-known neighbouring 798 arts district and 751 design park (Zhang 2016). The event was first launched in 2011 by Beijing Municipal Government, aiming at putting Beijing onto UNESCO's Creative Cities Network (interview with government representative).[1] Dashilar has been one of the event sites since the beginning by carrying out a project named 'New Landscape of Dashilar'. However, whole area of Dashilar has not participated in the Design Week since 2011, and until now, only a handful of hutongs in the area have directly engaged in the event. Among these hutongs, Yangmeizhuxiejie Street has been the focus of the project since 2012, and hence is one of the hutongs so far more substantially affected by the urban regeneration strategy. It will then serve as the example in this study to illustrate how regeneration is planned to be achieved in this conservation area.

Selling Yangmeizhuxiejie to the creative class and more

It is evident from the name of the project – 'New Landscape of Dashilar' – that the goal of participating in Beijing Design Week is to change the look of Dashilar and establish a new image of the area (see Chapter 16, this volume). The project now consists of two main components: first, houses that have already been emptied out through the 'micro-circulation' approach are filled in with pop-up shops and exhibitions; and second, since 2013, artists and designers are invited to propose ideas for improving the look and conditions of courtyard houses and streets with a chance of experimenting those ideas during Beijing Design Week. Both of these two components are temporary and pop-up in nature. While the first component is similar to many other pop-up uses for commercial or creative purposes, actions in the second component usually do not lead to long-term solutions or stay in place after the event either. Most of the 'interventions' offered by artists and designers are experimental and removed when the Design Week is over (see Figures 10.1 and 10.2). For example, on Yangmeizhuxiejie Street, many of these experiments are displays of innovative ways of using the street space and are removed afterwards to return the street to its everyday look and use.

The project 'New Landscape of Dashilar' turns Yangmeizhuxiejie into a place that is saleable to the creative class – both the producers and the consumers of 'creative products'. The attractiveness of Yangmeizhuxiejie to urban middle-class, creative-class consumers lies in its unique aesthetic. It is distinguished from other hutongs in Beijing's conservation areas, and especially the already well-known Nanluoguxiang, through the event-based pop-up landscape created during Beijing Design Week. On one hand, event-based pop-up uses assume an ephemeral but also spectacular nature, which interests those consumers who constantly search for novel experiences. And on the other hand, the process of selecting participants in the event and the framing around art and design make the uses that pop up on

FIGURES 10.1 and 10.2 A pop-up shop during and after Beijing Design Week

Source: Amy Yueming Zhang

FIGURE 10.3 A design idea experimented during Beijing Design Week

Source: Amy Yueming Zhang

Yangmeizhuxiejie during Beijing Design Week cater more to those who regard places like today's Nanluoguxiang as too popular and touristy (see Figure 10.4). Thus, Yangmeizhuxiejie acquires a saleable 'personality' (Smith 1996: 19) in a niche market, as a place that entails 'distinction' (see Bourdieu 1987) for those who visit.

FIGURE 10.4 Pop-up uses with an emphasis on book design during Beijing Design Week

Source: Amy Yueming Zhang

The appeal of Yangmeizhuxiejie to creative producers – artists and designers – is also the result of the event-based pop-up landscape. Through this mechanism, a prospective look and feel of the street is created, in the presence of potential customers. It is then possible for creative producers to envision having a long-term location on Yangmeizhuxiejie, where they can be neighbours with others who engage in similar activities and access a pool of customers relatively easily. Organisers of the 'New Landscape of Dashilar' project also mobilise the 'community' rhetoric mentioned above in the vision of Dashilar to attract this group of potential tenants. In their words, the future of Dashilar is to become a design community where creative producers gather to work and live and identify themselves as 'residents of Dashilar' (Dashilar 2013a). And to enhance this image of a close-knit community where people develop a place-based identity, a series of forums were conducted in the summer of 2016 for artists, designers and business owners who already choose long-term locations on Yangmeizhuxiejie to share their experiences as 'people of Yangmeizhu' (Dashilar 2016). As a result of the combination of pop-up during Beijing Design Week and the rhetoric of community, Yangmeizhuxiejie witnesses a gradual increase in the presence of commercial premises and studios (see Figures 10.5 and 10.6), which reinforces the brand it establishes through Beijing Design Week and its attraction to urban middle-class, creative class consumers.

FIGURE 10.5 A long-established gallery close to Yangmeizhuxiejie

Source: Amy Yueming Zhang

FIGURE 10.6 A construction site in Yangmeizhuxiejie (after Beijing Design Week in 2013) which became a popular restaurant

Source: Amy Yueming Zhang

Yangmeizhuxiejie is thus 'sold' to the creative class – both the producers and consumers of 'creative products' – as a 'place to be' with a distinct image. The same model has been applied to other hutongs in Dashilar since 2014, which shows the support of the district government to the approach. Despite the claim from government representatives regarding the intention of avoiding repeating the situation of Nanluoguxiang in Dashilar, there is no guarantee that hutongs like Yangmeizhuxiejie in this area will not face intense commercialisation and gentrification in the future once their brand is established and a saleable landscape is created (see Chapter 3, this volume). In fact, given that the goal of engaging with art, design and creative industries is to regenerate the whole area, the district government's focus is more on the rent level than on building a 'creative community'. As is stated on the website of Dashilar, the strategy is to first attract industries and uses that may not be able to afford very high rent but have the ability to bring other uses that can pay higher rent into the area (Dashilar 2013b).

The case of Dashilar thus involves two rounds of producing a saleable landscape: first, event-based pop-up uses create a landscape that is saleable to the producers and consumers of 'creative products'; and second, based on the scene established by these producers and consumers, a landscape that is saleable to investors and to potential tenants with ability to pay higher rent is produced. In these two steps, spaces used by art and design industries (in both pop-up and long-term formats) are first used as tokens to establish confidence with the area and attract similar uses, and then are designated as catalyst to facilitate regeneration and revalorisation in the area. However, so far it is still unclear whether and to what extent the second step is going to be achieved.

Conclusion

This chapter provides a brief account of an urban regeneration project in Beijing that employs 'creative' temporary/pop-up uses for creating a saleable landscape to the creative class and beyond (also see Chapters, 3, 7, 8 and 16, this volume). As I state above, given the pace and extent of the project, it is still too early to draw a conclusion on its effects, and more in-depth research is also needed for assessing this urban regeneration strategy. Therefore, this final section discusses a number of implications of this case-study and provides thoughts on possible directions for future research.

First, the case of Dashilar in Beijing can be viewed as one more example of the recent fashion of employing temporary/pop-up uses for triggering urban regeneration and property revalorisation. However, different from most cases studied in existing research where temporary/pop-up uses are engaged for changing the image of vacant, 'abandoned', and 'wasted' urban spaces (Colomb 2012; Harris 2015; O'Callaghan and Lawton 2015; Schaller and Guinand 2017), the project in Dashilar is built upon intentionally created vacancy. Spaces that are occupied by pop-up uses during Beijing Design Week and later by long-term 'creative' uses become available through 'micro-circulation', which empties out residential uses that were in place. This case thus provides materials for further discussions on how the relationship between urban vacancy and temporary/pop-up uses unfolds in various ways in different contexts, and the dynamic between the images of and the use, or lack thereof, of certain urban areas.

This strategy of employing temporary/pop-up uses in intentionally created vacant spaces for changing the image of an inner-city neighbourhood in Beijing is also a strategy that uses consumption-led regeneration to justify urban cleansing and displacement. In the phrasing of this urban regeneration project, the 'problem' of the Dashilar area is one of having 'outdated industry' that cannot meet the needs and taste of today's consumers, which requires a revitalisation plan that can bring new and creative industries (Dashilar 2011). This rhetoric is based on a questionable emphasis on catering to middle-class consumers over the needs of local residents; and secondly conceals the fact that the injection of creative industries into the area takes place in previously residential spaces, thus the ultimate goal of the project on urban cleansing and land revalorisation. The choice of mobilising this particular rhetoric demonstrates that consumption has become a powerful political discourse in China. Catering to and fulfilling the consumption preference of the urban middle class now almost assumes an automatic legitimacy and is regarded by the Chinese government as an effective justification for urban socio-spatial changes and as a tactic to co-opt the middle class into depoliticising such changes (see Chapter 8, this volume). Moreover, the replacement of residential use with commercial use, with the help of interim vacancy, shows the government's prioritisation of the nature of the same built environment as property over housing, and prioritisation of exchange value over use value.

Finally, this case resonates with a question that has been raised multiple times by previous research on using art and culture for urban regeneration, namely the complex and sometimes controversial role of the art and design community in the process. Whereas some point out that although artists as 'pioneers' could initiate a first round of gentrification, they often become victims of gentrification later and have to face displacement (Mathews 2008; Zukin 1989), others argue that the art community is complicit with gentrification, and 'to portray artists as the victims of gentrification is to mock the plight of the neighborhood's real victims' (Deutsche and Ryan 1984: 104; Smith 1996). The debate on whether and to what extent the art and design community is complicit with urban regeneration strategies in displacing residents and gentrifying neighbourhoods continues in research of such cases, and the case of Dashilar in Beijing provides another opportunity to study how artists, designers, and the 'creative class' perceive their roles and their relationships with the government in the process of urban regeneration.

Note

1 Beijing became a member of the Creative Cities Network in 2012 as a Creative City of Design (UNESCO, 2012).

PART IV

(Im)mobilities and materialities

11

URBAN CROSS-BORDER MOBILITIES

Geopolitical encounters and bordering practices of 'Taiwanese compatriots' in China

J. J. Zhang

Introduction

The China – Taiwan conundrum remains one of the unresolved conflicts of the Cold War era. Although it can be said that both political entities are relatively at peace with each other, no treaty has ever been signed, and China remains ardent that it will use military action against Taiwan should the latter proclaim independence.[1] However, the increase in Chinese political, economic and cultural power over the last few decades has led to the two republics engaging with each other in new politicised ways. Taiwan has increasingly come to terms that 'independence' is simply not a realistic option. Pushing for independence could only upset China and strain both cross-strait and international (US) relations. China, on the other hand, is beginning to abandon efforts in engaging Taiwan in non-constructive verbal disputes over sovereignty, in preference for the potential economic benefits of a 'Greater China' sphere of co-prosperity. Such sentiments for peaceful and mutual economic development are neatly captured in existing and emerging tourism activities in, and between, the two republics.

Although the normalisation of travel between the two former enemies is a welcome development, politics are never eradicated from seemingly banal activities, with everyday local realities often challenging the framework of 'peace through tourism'. Rather than seeing this as 'economics before politics', cross-strait engagement has metamorphosed into something that not only concentrates on macro-political issues, but micro-political nuances as well, and this has profound implications on how we study urban encounters on both sides of the Strait. This chapter thus theoretically interrogates the 'urban cross-border mobilities' of Taiwanese tourists in China. I move beyond mainstream urban studies' focus on specific cities, by examining tourists' travel experiences at border areas through the lens of 'materialities' and 'liminality'. I argue that 'edges of a city' are vital to understanding Chinese urbanism. These border areas are not empty spaces of transition, but are filled with identity negotiations and performances. This chapter also offers a critical reflection on encountering the 'other' – a typical urban condition (Valentine 2008), as cross-strait interactions unfold (also see Chapters 5, 7 and 14, this volume).

This chapter is divided into three parts. The first section focuses on the Taiwanese Compatriot Permit and the embodied experiences of border-crossing, examined through the viewpoint of 'liminality'. This is followed by an exploratory theoretical foray into the realm of 'play' and 'humour' associated with Taiwanese tourists' strategic negotiations with identity and travel documents. The final section highlights how geopolitical encounters and bordering practices are furnished by, and performed through, the trading of things between Taiwanese tourists and Chinese locals at a ferry terminal in Xiamen, China. I argue that we need to develop new critical perspectives to fully investigate the complexity of Chinese urbanism and here I highlight how '(im)mobilities', 'materialities', 'liminality' and 'play' allow timely and important theoretical and empirical reflections on Chinese cities in an increasingly interconnected urban world.

Materialities and liminality in urban cross-border mobilities

Relational perspectives in urban theory focus on 'flows, movements and connections' instead of 'territories, scales and boundedness', thereby regarding the city as constantly in the processes of becoming and reordering (Söderström 2017: 197). It can be said that the social sciences, particularly urban studies, have experienced a 'mobility turn' (Cresswell 2010), as there is an increasing wealth of literature on places and practices of urban mobility as well as methods of studying mobilities (Söderström 2017). A notable change in perspectives, however, pays attention to the 'embodied, interactive and emotional experience of mobility' rather than its 'molecular or atomist visions' (ibid.: 198). As Cresswell (2010: 19) reminds us mobility is constituted by physical movement, representations and practices and this chapter primarily focuses on the last part, i.e. the 'experienced and embodied practices of movement', which has oftentimes been overlooked by traditional mobility studies including transport geography and migration theory. Studying cross-border mobilities from a grounded approach will contribute to understanding urban (im)mobilities as experienced by ordinary people (see Chapters 12 and 13, this volume).

One way to gain a more intimate sense of cross-strait tourists' experiences is via materialities; that is, to look at the emotional transactions between people and things that are associated with their travels, and how these are in turn connected to wider geopolitical contexts. For example, travel documents that are close to the personal or those that are part and parcel of a touring experience are far from inert; they participate in the social and political lives of their owners, feature in bordering practices between the Chinese and the Taiwanese, and are often platforms through which identities are performed (see Chapter 9, this volume). Furthermore, things matter to people, but they do not merely serve as an 'extension of self' (Belk

1988), that is, 'what one is'; things also contribute to 'how one is' (Sayer 2011). In other words, tourists are suspended amongst other things during their travels and these things are capable of affecting their feelings, emotions and values. As such, in response to calls for new experimentations with potentialities of materiality (Anderson and Tolia-Kelly 2004), I hope to garner a more intimate understanding of Taiwanese tourists' travel experiences through things that are close to the personal and the everyday.

I am also interested in how tourists behave during their tour, especially at border-crossings or border areas (e.g. immigration checkpoints). I suggest that the concept of 'liminality', famously developed by Arnold van Gennep and later by Victor Turner (1969), provides pertinent insights. According to Turner (1979: 465), 'liminality' literally means 'being-on-a-threshold' – 'a state or process which is betwixt-and-between the normal, day-to-day cultural and social states and processes of getting and spending, preserving law and order, and registering structural status'. These in-between places constitute a liminal space within which normativities of the tourists' everyday lives are temporarily kept in suspension, allowing them to encounter the 'Other' in a different social structure. Utilisation of 'liminality' in tourist/tourism studies is not new. A quick reference to existing literature shows the concept being applied to society's/individuals' behaviour, activities (e.g. sex tourism, pilgrimage, etc.), and specific sites/places (e.g. hotel). For instance, Wagner (1977) adapts Turner's notion of 'communitas' and argues that tourists form 'spontaneous communitas' and interact with each other based on 'the spirit of the holiday' rather than 'the home life social hierarchical system' (see Chapters 5, 7, 12 and 14, this volume). Gottlieb (1982) on the other hand, experiments with the inversion of the everyday identities of holiday-seekers: the upper-class tourists temporarily becoming a 'pseudo-proletariat', while the middle-class ones seek an aristocratic change when on tour. Building on this genre of 'inversionary behaviour', Lett (1983) incorporates the concept of 'play' as developed by Huizinga (1950, cited in Currie 1997) and Norbeck (1971, cited in Currie 1997) to explicate yacht tourists' sexual behaviour. Play, according to Stevens (2017: 223), 'is an emancipation from the routines, constraints and preconceptions of everyday social existence', and expresses 'social ideals that are difficult to achieve in everyday life'. Tourism, for Lett, is a form of play, 'a stepping out of "real" life into a temporary sphere of activity with a disposition all its own' (Huizinga, cited in Lett 1983: 41). Pritchard and Morgan (2006) bring this discussion into the hotels, seeing them as 'liminal sites of transition and transgression'. As is evident, the concept of liminality has been well adapted in studies on the socio-cultural aspects of tourism especially in the realm of sexual activities. The geo-political potential of it seems to be under-theorised. Although scholars like Salter (2003) and Wang (2004) alluded to the 'rites of passage' of passport checks and the humiliating experiences of travellers as they undergo rigorous scrutiny by the immigration officers, the existential inner-workings of the travellers during such a liminal period has yet to be explored.

Rapprochement tourism between China and Taiwan offers a fertile context to explore these geo-political implications of 'liminality'. The 'political' may refer to both the macro-politics of cross-strait relations and micro-political practices of the tourists, while the 'geo' represents the places, the in-between, the marginal or the transitional, where encounters amongst people and things happen. Having a better appreciation of the 'material moments' (Burrell 2008) during this liminal period may have significant implications for developing a deeper understanding of rapprochement tourism between politically divided entities. The rest of the chapter looks at how such interaction presents itself in a series of travel narratives by Taiwanese tourists. These stories might be subjective and personal, 'yet they are not just free-floating "values" or expressions projected onto the world but feelings about various events and circumstances that aren't merely subjective' (Sayer 2011: 1).

Crossing the border: checkpoints, travel documents and the performance of identity

One of the recurring themes in my research with Taiwanese and Chinese tourists on their travel experiences is that of their interactions with personal documents like passports and entry permits.[2] Such travel documents, the applying for them, the possession of them, and their usage often evoke affective moments at border crossings. I suggest that by listening to their travel narratives and plotting the moments when their travel documents remind them (and others) of their identities and in the process facilitate or impair their mobility we can gain a better understanding of cross-strait relations at a more intimate level.

For Taiwanese tourists travelling overseas, the Taiwan (Republic of China – ROC) passport (see Figure 11.1) can prove to be a hindrance. Although it can be said that Taiwan has *de facto* independence, it is not recognised as a sovereign state by the United Nations. Both China and Taiwan of course claim to be the 'true China'. The China – Taiwan conflict has never been resolved but a compromise was reached under the somewhat ambiguous 1992 Consensus, where both China and Taiwan confirm that there is only 'one China' albeit each having a different notion of what that 'China' is. However, most countries adopt a 'one China policy', recognising the People's Republic of China (PRC) as the official China. The validity of the Taiwan passport is thus questioned at immigration checkpoints and has impeded mobility rather than facilitated flow.

Needless to say, the Taiwan passport is not recognised by the Chinese authorities, and vice versa. Even in the context of rapprochement tourism, both sides are cautious not to grant each other *de jure* sovereignty. As such, instead of stamping the passport directly, tourists from either side are issued travel permits to be shown to immigration authorities upon arrival. A Taiwanese tourist will be issued the Mainland Travel Permit for Taiwan Residents (see Figure 11.1), commonly referred to as the Taiwan Compatriot Permit (TCP), while the Chinese counterpart will be issued the Exit and Entry Permit for the Taiwan Area of the Republic of China. The TCP has been in existence since 1987 when the then Taiwan President, Chiang Ching-kuo, lifted the travel ban across the Taiwan Strait. A typical Taiwanese visiting mainland China would use the Taiwan passport at the Taiwanese checkpoint,

FIGURE 11.1 The Taiwan Passport (*left*) and the Taiwan Compatriot Permit (*right*)

Source: J. J. Zhang

and present the TCP to the Chinese authorities upon arrival. A TCP holder is identified as someone who resides in Taiwan, but who is essentially a fellow countryman hence 'compatriot'. I am interested in how the possession of the TCP affectively interacts with the Taiwanese tourist at the Chinese checkpoint and how the checkpoint itself creates a liminal environment between home and away. One of my interviewees, Ben, who has recently been to Shanghai, offered his 'experience of mobility':

B: When my friend and I reached the border control at Shanghai airport, I actually wanted to queue at the 'International Arrival line' … I asked, 'Why should we queue at the "Domestic Arrival line"?'

J: So your friend naturally went to the Domestic queue?

B: No, that's because he had travelled to Shanghai many times, he said we had to go to the Domestic line. Then I asked, 'Shouldn't we go to the International one?' He answered, 'No, we have to go to the Domestic line.'

B: As for the TCP … I got hold of this thing as I need to go to Shanghai …I treated it as merely a document.

J: Just like a visa?

B: Yes … and I have nothing against it. But I would think about where to queue at the immigration. Why should we queue at the domestic line if people from

Hong Kong and Macau are allowed to use the international one? Hong Kong and Macau … they are China, not us.

But in the end, I still went to the Domestic line so as not to get into trouble. It's different in China. In Taiwan, you can scold the government, but in China, you need to be more reserved.

Ben's recollection of his border experience was infused with politics and with questions about struggles of identity and mobility. As Ben's narration clearly explicates, the material significance of his TCP and the identity it represents do not correspond to the holder's self-identification. Yet the affective moments between Ben's TCP and himself did not happen before the traveller reached the Chinese checkpoint. The problematisation of where to queue shows that he was well aware of the identity politics of cross-strait tourists, and reflects the identity struggle he had to engage with. To be queuing in the domestic line, was to admit that Taiwan is part of China. Choosing the international line, however, was a performance of his national identity as Taiwanese. However, this struggle and resistance was short-lived as his travel companion, who had more experience in crossing the border, advised him against it. Overwhelmed by the stringent mobility regime and fear of 'get[ting] into trouble', in a place that was neither here nor there, he decided to follow the 'rules' and joined the domestic arrival line.

These political overtones as something that is as 'personal' as a passport or a travel permit need to be interrogated further. In contrast to the feeling of humiliation reported by Wang (2004) of Taiwanese travellers when their passports were deemed invalid at international checkpoints, my informants did not allude to any form of 'embarrassment' when using the TCP. Indeed, it was 'only a document'. I suggest that at liminal spaces like immigration checkpoints, tourists tend to keep their values or beliefs in suspension; the 'tourist identity' allows them to be in a state of political numbness, to compromise to the institutional requirements, even to the extent of using a travel document that recognises indirectly that Taiwan is part of China.

Let us return to the Chinese checkpoint in order to interrogate in more detail Taiwanese tourists' experiences beyond queuing at the domestic arrival line. One of my interviewees, Chen, gave an interesting account of her psychological and 'bodily' 'transformations' when crossing the border:

C: At the immigration point, you are already in mainland China, but you are not really in the country … Whenever I reached the place, it seems like I am changing to another person. Because I will always imagine that I was about to enter an uncivilised place … a place where people spit freely everywhere … haha … This is what we imagine Chinese people to be. And their low level of civilisation … how they always jump the queue, and push their way around … I was about to enter such a place … So whenever I was there, I would start to think of myself as becoming another person.

J: To be like them?

C: No! Haha … Just that I have to pretend to be detached/unconnected/indifferent. You'll start to feel that you ought to equip yourself with arms and armour.
J: Oh I see … to protect yourself.
C: Yes …
J: Because you felt uneasy?
C: Hmm … No. Just act nonchalant … Moreover, Taiwanese are always being cheated in China. I felt that we are like idiots. My friends and I were cheated several times and the feeling wasn't good. For example, the porters will ask for a certain fee and later charged higher. So once you reach the immigration point, you have to start to be a little different … you'll have that feeling.

The Chinese checkpoint is indeed a threshold for Chen. She felt that she was not in Taiwan, but not quite yet in China. This liminal space created a kind of anxiousness in her while she prepared to encounter the other side. It was a transitional period during which her social imaginations of China and its people materialised in her 'transformation' to become 'another person'. Chen's narration of how she had to transform herself before officially entering China was akin to changing into a 'tourist mode' (Currie 1997). This was not any ordinary mode, but one in which she became 'detached/unconnected/indifferent' and equipped with 'arms and armour'. Chen's border crossing experience reminds one of the 'rites of passage' that act as both 'indicators and vehicles of transition from one sociocultural state and status to another' (Turner 1979: 466). To illustrate, Van Gennep introduced the three phases of:

> (1) separation (from ordinary social life); (2) margin or limen (meaning threshold), when the subjects of ritual fall into a limbo between their past and present modes of daily existence; and (3) re-aggregation, when they are ritually returned to secular or mundane life – either at a higher status level or in an altered state of consciousness or social being
>
> *(Turner 1979: 466 – 467).*

Juxtaposed onto the border crossing process, Chen left her ordinary life and everyday social structure when she embarked on her travel to China. She then entered the liminal period at the immigration checkpoint where she felt that she was becoming another person. Finally, when she was granted entry and crossed the checkpoint, she would be in a different socio-cultural state as she began her journey under a new social structure. Her border crossing experience therefore appears not dissimilar to Gennep's 'rites of passage'. From this evidence, it is clear that tourists' experience of liminality are not straightforward during or even after the checkpoint. Notions of identity performance and behaviour, and preparations for the encounter with the 'other' continue to be worked through while waiting for or going through passport/travel permit inspection. These emotional and embodied identity performances contribute to the '"furnishing" of journey and border times and spaces' (Burrell 2008: 353), and help us better understand the mentality embedded in urban encounters.

Play and humour: Taiwanese identification cards and identities

Discussion hitherto has focused on aspects of self-identification or identification by others. Here, however, I suggest that in the context of rapprochement tourism between China and Taiwan, there lies an element of 'play' in such identity performances. In this section, I draw on the usage of identification documents other than passport or travel permits by Taiwanese tourists in China, and the circumstances in which play and humour are utilised by respondents as a way to straddle identity boundaries. For the Taiwanese tourists, these documents are not necessarily travel documents *per se*, but enable advantageous negotiation of macro-political structures. One of my Taiwanese respondents told me about her friend's routine performance of identity to Chinese locals:

> My friend is very funny. Whenever he goes to China, he would show off his Republic of China [Taiwan] ID card to people there and quipped, 'We belong to the same country right? In that case, I should be able to use this here!' Haha … He just felt that with his effort, he could influence the Chinese people to think that Taiwan is a sovereign country. He's always like that and ended up in long and funny debates with shop owners. We'll just leave him alone to enjoy himself.

Stevens (2017: 219) refers to 'play' as 'a low-risk means for all people to experiment with social practices, roles and rules'. As Turner (1979: 466) also observes, 'Liminality is full of potency and potentiality. It may also be full of experiment and play. There may be a play of ideas, a play of words, a play of symbols, a play of metaphors. In it, play's the thing.' Such playfulness can be extended to this tourist's identity performance through his Taiwanese Identity Card. Although Taiwanese tourists in China are required to use their Taiwan Compatriot Permit (TCP) for identification purposes, he chose to do otherwise. In the liminal state of touring, he put forward his views on Taiwanese sovereignty in a playful and light-hearted manner that might not be possible under the host country's normal social structure, thereby 'testing boundaries' (Stevens 2017: 226). He would not go so far as to mock the Chinese state, but still found liberties in the way he expressed himself and interacted with the locals. Conversely, in the spirit of tourism, the locals were happy to engage him in friendly debates on cross-strait relations.

Although the connections between liminality and the notion of 'play' have been discussed in tourism studies, research on the relationship between a closely related concept – humour – and tourism (Pearce 2009; Pearce and Pabel 2015) seems to overlook the efficacy of the liminal. In his discussion on how humour is utilised in place-marketing, during the tourists' on-site experience, or by ethnic groups while performing for tourists, Pearce (2009) does not pay particular attention to the embodied and emotional experience of the tourists. I argue that beyond seeing humour as something 'for tourists', 'about tourists' or 'by tourists' (ibid.), we should also explore the liminal conditions that afford and affect how humour is being delivered and received.

To illustrate. Taiwanese tourists seem to avert the political implications of using the TCP in a playful and humorous manner. One of my respondents quipped, 'Do you know what we [Taiwanese] call the TCP?' I did not. 'Dai bao zheng,' she replied ('Permit for idiots'), before bursting into laughter. The interesting play with words and pronunciation (from 'tai bao zheng' to 'dai bao zheng' – a near-homophone to 'tai bao zheng') represents a self-humiliation but meanwhile a way to avoid politically sensitive sovereignty issues. As idiots, they do not need to fuss over issues of national identity; their objective is simply crossing the border to the other side. This resonates with Jansen's (2009: 820) recollection of his conversation with a Serbian woman who lamented that the visa for Serbians to enter Hungary is widely referred to as 'Porez za budale' – 'Tax for idiots'. For Taiwanese tourists, it seems that people have come to terms with their country's ambiguous identity (see Corcuff 2012 for a discussion on the liminality of Taiwan from a geo-political perspective) and instead of being offended by the visa regime as suggested by Wang (2004), there was a denouncement/deprecation of the self in a playful yet politically informed manner. Such humour does not connote a foolish or naïve gesture, but I argue is a kind of 'soft' yet powerful resistance to the institution.

Furthermore, Taiwanese tourists I spoke with sometimes boasted of their triumphs over the 'system' when their Taiwanese identification documents were accepted by Chinese authorities. This is an extract from my interview with Lin, a Taiwanese tour guide:

L: Whenever I go to China, I'll bring my tour guide license along. In China you need to pay to enter most tourist attractions. But tour guides enter for free or are at least given huge discounts when you show your tour guide license.

J: They recognised the Taiwanese tour guide license?

L: Not really. Some attractions do acknowledge, but others don't.
There was once we went to a scenic spot at Nanjing. The entry fee itself was already CNY￥120. Fortunately I've got my tour guide license with me. I depended on it to help me save lots of money!

J: So what were you thinking when using the Taiwanese license on Chinese soil?

L: Mine was very simple: to save money! Hahaha. I don't care about national identity. It's all about saving money.

J: So when you were using the pass, what did you actually want the person at the ticketing counter to think?

L: Hahaha …very interesting! Of course I want the person to think that 'ok you can use this, you are a tour guide'. And I will presume that the person was thinking, 'Good, you acknowledge that Taiwan belongs to us'… Hahaha … If they don't accept, it means that they don't recognise that Taiwan belongs to them. It will be interesting if I were to probe further in such a situation, but normally if they don't accept my Taiwanese document, I'll just take it back, because I'm in a foreign place.

J: So it's a win-win situation no matter what. If they accept the pass, they are happy, because they will feel that you think Taiwan belongs to China. On the other hand, you'll be happy as you have saved some money.

L: Haha … but you'll feel that you've done something wrong … hahaha …

J: If they don't, it means that they do not recognise that Taiwan is part of China. In that case, you'll be happy too!

L: Haha … Taiwan is not part of them at all … just that we wanted to take advantage! If they don't accept, we can say 'hmm … I think there's some problem with your notion of national identity'… hahaha … this method is great … I'll have to use it next time!

The producing of Taiwanese documents at ticketing counters of Chinese scenic spots is also quintessential example of 'play' in a liminal period of travelling. Lin claimed that it has got nothing to do with national identity as she was very clear about her political allegiance; since whether or not to 'play' is largely dependent on the wider social context and the expected responses of others (Stevens 2017), she knowingly exploited the ambiguity of the macro-political notion of 'belonging to China'. The usage of the Taiwanese document was interpreted as a submission to the sovereignty of the People's Republic of China (PRC). Indeed, documents 'take the shape of or transform into affect and become part of their handlers in that way' (2007, cited in Jansen 2009: 816). Furthermore, this interview extract is interesting as it not only reveals the tourist's material moments with her tour guide pass in China, but also hints at her enthusiasm in experimenting with new 'techniques' to take advantage of the current political climate in cross-strait relations. It shows that such performance of identity is not just talked about in the aftermath of travel, but is constantly in the making, constantly becoming scripted in the tourist's mind, and constantly being 'humourised'. This performance would be rehearsed and staged in future encounters with the hosts in China. Butler (1993, cited in Sofaer 2007: 3–4) reminds us that 'people can hold multiple or plural identities which may spring to the fore in different circumstances, times, and places'. In this sense, the allegiance of Lin to the PRC was derived based on a double assumption. Lin assumed that by producing her Taiwanese tour guide pass, the Chinese authority would assume that she was recognising and submitting herself to the PRC sovereignty, and would then grant her a discount. As such, her clear allegiance to the Republic of China (Taiwan) co-exists with her willingness to be assumed as having a sense of belonging to the PRC. However, as is evident from the conversation, Lin might be haunted morally when she returned to her home social structure as she felt that she has 'done something wrong' in betraying her own national identity. Yet, in the spirit of tourism, at the liminal stage where she 'plays' with her identity and the other's identification of her, the ultimate aim was not about making macro-political statements, but a personal triumph of saving money. This economic rationale in the political life of things might seem trivial but nevertheless plays an important part in Taiwanese tourists' mobility experience in cross-border urban areas. The final section of this chapter seeks to extend this economic *raison d'être* to the analysis of tourists' buying and selling of things at the border.

Street economies: buying and selling at a ferry terminal

This final substantive section of the chapter explores how an urban area of transit for cross-border tourists is not just an empty place – it is not 'placeless' – but is furnished with complex and heterogeneous materialities and discursive construction of identities. More specifically, I focus on the Dong Du Ferry Terminal in Xiamen, China as an example (see Figure 11.2). During my fieldwork in China and Taiwan, I was quite often based in the Taiwan-held island of Kinmen. Although not the political centre of Taiwan, this island, situated in between the two 'Chinas', is very much at the heart of cross-strait politics. Even before the commencement of direct flights between China and Taiwan, Kinmen had been used as a test-bed for re-establishing links between people from both sides. What was called the 'mini-three links' saw daily ferry services between Kinmen and Xiamen in Mainland China. It did not take long for me to realise that once the Taiwanese tourists reached the Xiamen ferry terminal they would be approached by locals – often attendants from souvenir shops – asking if they have any duty-free cigarettes or liquor to sell. Taiwanese tourists would usually spend about fifteen minutes at the arrival hall of the ferry terminal waiting for their coaches and would often participate in informal street economies (see Chapter 13, this volume). The tourists I spoke to revealed that the profit gained from the selling of duty-free items, which they have bought in Kinmen, would be used as 'pocket money' for their tour or spent at the very shop they sold their 'goods' to.

FIGURE 11.2 Taiwanese tourists at the arrival lounge of the Dong Du Ferry Terminal

Source: J. J. Zhang

Interestingly, these shops are not merely souvenir shops, but sell an amazing range of counterfeit mobile phones, including the ever-popular iPhones. I was curious about such 'street economies' and interested in how tourists and locals interact, bargain and trade during this space-time at the terminal. Fortunately I was able to follow a friend, Ren, from Kinmen, who had decided to spend the weekend in Xiamen and had plans to earn some pocket money by selling duty-free items at the ferry terminal. Before the trip, I had a conversation with Ren:

R: There is an economy out there at the ferry terminal. Normally, we will bring 2 boxes of cigarettes … like Mild Seven … We can each bring a bottle of alcohol as well. The first choice is western liquor like Martell Blue. If that's sold out, we usually go for the Kinmen Kaoliang Liquor, which will earn us less profit, but better than nothing. So Mild Seven and Martell would probably earn you a profit of CNY￥240. A Mild Seven and Kaoliang combination will fetch about CNY￥110. CNY￥240 … that's already three quarters of my return ferry ticket!

J: Is this legal?

R: I'm not sure, but I don't think the authorities can say anything. These're just small and private transactions between people. But there is some sort of regulation that you can't bring more than one bottle of Martell across the border per month. Sometimes the Chinese customs stops and checks you.

J: Other than selling these duty-free goods, do you buy things from the local shops at the ferry terminal?

R: Yes, of course … especially mobile phones. I've bought a couple from them in the past and they work well.

As he had earlier decided to buy a counterfeit iPhone at the Xiamen terminal, he took duty free Mild-seven cigarettes and Martell liquor directly to the 'phone' shop, hoping to strike a deal with the shop owner. The negotiation for the price of the duty-free goods went well and he earned a handsome profit of CNY￥240. However, it was during the exchange of words between Ren and the shop owner when he was bargaining for a better price for the counterfeit iPhone that notions of identity and personal boundaries started to unfold:

O: CNY￥900 is the best price. You guys are from Taiwan … I have relatives there too, that's why I'm giving you the best price!

R: Come on … CNY￥800. We *are* compatriots right? Compatriots should be nice to each other …

O: CNY￥890 is the lowest I can go. You are not Taiwanese! Taiwanese people are more generous and fun to do business with!

R: Well, we are not Taiwanese you see. We are Kinmenese!

By associating with his Taiwanese relatives, the shop owner was attempting to dismantle the personal border, based on nationality, between him and his Taiwanese customer. He was also trying his best to speak Mandarin in a Taiwanese accent so as to break any possible language barrier. Conversely, pushing for a good deal for the counterfeit iPhone, Ren took 'tactical advantage of the current system' (Wang 2004) and rapprochement climate by highlighting his 'compatriot' status. In such an economic interaction at the ferry terminal, personal boundaries were withdrawn as buyer and dealer negotiate their identities in order to seal a good deal. But as soon as the shop owner realised that his tactics were not working as intended, he immediately withdrew and put up a 'wall', claiming that his customer was not 'Taiwanese'. The buying and selling of things at the ferry terminal shows how places of transition 'are intensely material and used spaces and times' (Burrell 2008: 357). As the lounge at the ferry terminal constitutes a space-time where tourists are expected to be waiting for either their tour coaches or ferries, the flow of tourists in such an in-between space is not a priority (compared to, for example, the immigration checkpoint) and this creates opportunities for interactions between the Taiwanese tourists and Chinese traders. Indeed, at the waiting lounge, one experiences a transition from a formal space of security and efficiency to one of informalities and street economies. Instead of an empty 'in-between space, suspended between two realities' (Burrell 2008: 369), the ferry terminal is brought to life by the bordering practices and economic transactions, which create unique spaces of encounters, and point to implications for further studying urban mobilities, as well as significant socio-cultural impacts.

Conclusion

By bringing 'materialities' and 'liminality' to the study of (im)mobilities in an urban setting, this chapter has provided a grounded theoretical and empirical understanding of cross-strait tourism between China and Taiwan. From the various tourist travel narratives, I have attempted to locate and highlight the confluence of 'the political and personal' at both the real and imagined borders. As is explicated throughout my discussion, the ubiquitous border does not exist only in its physical form; imagined and perceived borders are equally potent in (re)shaping cross-strait relations. Furthermore, it is evident from the discussion that material things close to the personal or those that are part and parcel of a touring experience are far from inert; they participate in the social and political lives of their owners, and are often platforms that connect 'macro structures (the state) and micro actors (individuals) to each other' (Wang 2004: 355). Things like travel permits also engage in affective material moments with their holders at border crossings, influencing how they see themselves and how they perceive others' identification of them. Other than travel permits, personal documents like the Taiwanese tour guide licence and identity card were also explored. Through these examples, I elucidate that there is an element of 'play' in Taiwanese tourists' negotiation and performance of identities during the liminal period of travel. Their recollection and narration of the mobility experiences could be filled with humour and are at times self-deprecating. In addition, contrary to traditional analysis, the identity of a tourist during this liminal period is not merely inverted. Rather, I suggest a kind of diversion at play as s/he chooses to perform different identities at different social settings. Indeed, liminality provides a fertile conceptual ground to explore and explain the behaviour of cross-strait tourists, and opens up potential trajectories of how cross-strait relations may develop. As such, rapprochement tourism between China and Taiwan is not merely a political rhetoric, but something that is experienced at the personal level. However, it seems that in engaging with ephemeral or ubiquitous imaginations of the border, the significance of the physical border might be overlooked. In other words, discussions on border crossings in their own right might be seen as *passé* and thus not academically rigorous. In this respect, the identity negotiations undertaken by Taiwanese tourists at the Chinese immigration checkpoint, and the petty trading taking place at Xiamen Dong Du ferry terminal illustrate how in-between places are not empty, but charged with vitality and emotions.

In this chapter, I have also highlighted the ways in which research on Chinese urbanism can benefit from a critical engagement with the cultural geo-politics of cross-strait tourism, especially in the context of increasing global interconnection, both in urban China and beyond. By exploring 'the edge of cities', and focusing on encounters with the 'other', I have added value to broader theoretical debates regarding Chinese urbanism. Indeed, by asking ourselves: how can an engagement with concepts like '(im)mobilities', 'materialities' and 'liminality' inspire further epistemological and ontological re-theorising of the 'urban' and for that matter, 'Chinese urbanism'? How can we make better sense of urban encounters and interconnections from a geo-political standpoint? In seeking to address such questions, in this chapter I have highlighted the fruitfulness of many lines of inquiry that are deserving of sustained future research. My research on Taiwanese tourists' cross-border (im)mobilities in urban China with a geo-political twist show the importance of interrogating cross-strait relations not merely as residing in the realm of macro-politics, but as being permeated with material objects, emotions, identity negotiations performances and so on.

Notes

1 This chapter includes material published in Zhang, J.J. (2013) 'Borders on the move: cross-strait tourists' material moments on 'the other side' in the midst of rapprochement between China and Taiwan', *Geoforum*, 48: 94 – 101.

2 The research presented in this paper originates from a project on the cultural geo-politics of rapprochement tourism between China and Taiwan, and is based on ethnographic field research undertaken in 2010 and 2011. Seventy-seven semi-structured in-depth interviews with cross-strait tourists (forty-two Chinese and thirty-five Taiwanese) and participant observations were conducted over the one-year period. For this Chapter, I concentrate on Taiwanese travel narratives that highlight cross-strait tourists' material moments in order to gain a more nuanced understanding of cross-strait relations as experienced 'on the move'. The data collected and the interpretations made were context-based and the result of 'situated knowledges' (Haraway 1991, cited in Nightingale 2003).

12

CONTESTED (IM)MOBILITIES AND RHYTHMS OF CHINESE CITIES

Urban transformation and 'slow life' in Sanya

Jingfu Chen

Introduction

Chinese cities have undergone tremendous transformation spurred on by the state-oriented economic reform since the late 1970s. The past decades have witnessed the rapid reconfiguration of socio-economic structures and the formation of new urban spatialities (Ma and Wu 2005; Wu and Gaubatz 2012). Not only are material living conditions and urban environments substantially improved, but individuals' everyday practices and experiences are also continuously shaped and changed by new ideas, consumption culture and lifestyles, among others (see Chapters 2, 8 and 15, this volume). The rapid urban transformation of Chinese cities has also been stimulated by intensified inter-regional and global connections as well as increasing mobilities of people, products, information, data and 'things'. The dramatic and profound reconstruction of urban rhythms thus provides an important lens to understand the dynamic and complex nature of Chinese cities. However, in contrast to scholars' considerable interest in changing societies of metropolises like Beijing and Shanghai, limited attention has been given to the reproduction of rhythmic life in a small Chinese cities. The lack of research interest in small cities resembles a parochial perspective of urban theory, which relies too heavily on urban experiences of a small number of paradigmatic cities (Amin and Graham 1997; Bell and Jayne 2009; Robinson 2006).

This chapter thus examines the transformation of small cities in China by analysing the polyrhythmic urban life of Sanya, an island city in the South China Sea. I also contribute to advancing the 'new mobilities paradigm' (Hannam *et al.* 2006; Sheller and Urry 2006) by focusing on disparities between rapid urban change and traditional slow-paced lifestyles in local communities. Critical attention is directed to the production of complex urban rhythms through (im)mobilities as well as unpacking the different meanings assigned to 'fast' and 'slow' rhythms. My study of

Sanya offers insights into uneven urban transformation and the regulation of everyday life through discourses of modernity and development in China. Research data was collected during 2012 – 2013 and underpinned by critical (re)consideration of the polyrhythms of Sanya (see Lefebvre 2004).

The chapter begins with theoretical reflection of urban rhythms, (im)mobilities and meanings. This is followed by analysis of changing urban rhythms of Sanya with a backdrop of rapid socio-economic change in the past few decades. In particular, I highlight increasing velocity, seasonality and complexity of everyday routines and activities relating to transportation and tourism (im)mobilities. The subsequent section then focuses on everyday practices and contested meanings of slow-life amongst local communities, as illustrated by *Wuxiu* (afternoon snooze) and tea-drinking practices. In the context of the dramatic socio-economic reconfiguration and modernisation in Sanya, such traditional ways of living appear 'out of step' and are often considered as problematic. This chapter concludes with critical perspectives on modernity, changing societies and mobile life in small cities in contemporary China.

Urban rhythms, (im)mobilities and meanings

Rhythms are prevalent and characteristic of tempo-spatial practices. As Lefebvre argues, '[e]verywhere where there is interaction between a place, a time and an expenditure of energy, there is *rhythm*' (2004: 15 original emphasis). For Lefebvre, rhythm is repetition and interrelation between linear and cyclical procedures, which have specific circles of beginning, growth, recession and end. The feature of repetition does not indicate that rhythm is a homogenous reiteration. Indeed, Lefebvre contends that rhythm is distinguished from the metronomic and is productive of difference. Take everyday routines for example. They are not absolute and abstract repetition but embodied and social practices that are open for mutation and creativity (Gardiner 2000; Roberts 2006). Rhythms are inherently multiple and differentiated (see Chapter 14, this volume). Regarding linear and cyclical repetition, the former is related to monotonous social practices regulated by certain structures, while the latter comes from the cosmic and nature, and is illustrated by days, nights and seasons (Lefebvre 2004). The distinction and unity of the linear and the cyclical characterise various circles of complex rhythms.

The city is rhythmic and inter-connected time-spaces. Crang (2001), for example, argues that rhythms of the city are integrated with multiple temporalities and include transient affinities and integrations of various rhythmic activities and overlapped everyday routines. They are characterised by a wide range of flows, mobilities and practices of human and non-human agents, and manifest the dynamics of place (also see Edensor 2010). Smith and Hetherington (2013: 6) further contend that urban rhythms are productive of 'specific forms, configurations and relations of space, time, interaction and mobility'. Polyrhythms of the city are experienced and interpreted in different ways by individuals, creating distinctive senses of place. Lefebvre (2004) develops the idea of 'rhythmanalysis' to explore the dynamic and complex nature of urban societies through a wide range of urban rhythms. This approach emphasises everyday living and the experience of rhythmic cities. Similarly, Edensor and Holloway (2008) assert that rhythms circulating in and passing through spaces and places embody our ways of being in the world. Rhythmanalysis also examines how certain rhythms are regulated and normalised by socio-spatial structures (Lefebvre 2004). In line with these arguments, I discuss changing urban rhythms in Sanya and dominant discourses of fast rhythms and change.

In the modern era, rhythms of the city are subject to continuous transformation and reconfiguration through various (im)mobilities, which are fundamental to urban life (see Appadurai 1996). The history of modernisation and globalisation has led to increasing flows and movement of human subjects, information, capital and objects, with many people and areas linked closely together (Cresswell 2006a; Harvey 1989b; Thrift 2004). Massey (1995) explains that mobilities are characteristic of relational socio-spatialities. For example, with improved technology and transportation, people can make instant and synchronic communication with someone at a distance and develop new forms of social relations (Urry 2007). Against the background of intensified regional and international connections and mobilities, one direct consequence is speeded-up urban rhythms, everyday life and social change.

While such arguments are persuasive, we must not be seduced by the idea that everyone perceives and experiences accelerated rhythms and mobilities of the city in the same way. Instead, mobilities are differently performed and exert distinct influences on everyday routines and practices. While some people keep up with the increasing rhythms of place, others are 'out of step', and for some this suggests being 'left behind'. It is not uncommon however, to find significant inconsistencies between personal routines and rhythms of changing urban spaces. For example, according to Lefebvre (2004), arrhythmia refers to discordant rhythms and pathological conditions, and contrasts to eurhythmia, which indicates the harmony and integration of multiple rhythms and is viewed as 'normal' everydayness. The arrhythmia of the city can cause disturbance to individual life and urban transformation. In many cases, the disparity of individual and urban rhythms and people's different rhythmic mobilities are not neutral but shaped by various power networks (Adey 2010; Massey 1993). Cresswell (2010) thus argues that differences in motivation, velocity, rhythm, routes, experience and friction of mobilities are power-laden. For example, uneven urban rhythms may be caused by individuals' unequal rights to, and access to (im)mobilities, which are related to broader social inequalities and power relations.

Differences of rhythms and (im)mobilities are also further demonstrated by contested meanings given to rhythms of individuals' life and the city. In accordance with the 'new paradigm', mobility is far more than physical displacement but relates to socially constructed practices, which are constantly shaped by various ideas, values and discourses (Adey 2010; Cresswell 2006a; Urry 2007). Representations of rhythmic practices in mobile life are thus closely linked to dominant discourses and interpreted in different ways by individuals and social groups. Molz (2009) contends that change and speed has been heavily emphasised in modern 'Western' culture. This is also true of 'non-Western' societies. The ideology of modernity values transformation, efficiency and innovation at the expense of tradition and stagnation. Nomadism, speed and mobilities are conceived as progressive, productive and successful, while slowness and stillness are considered passive, inert and unable (Cresswell 2006b). Such discourses are illustrated in the contested meaning assigned to the traditional 'slow life' in Sanya.

Such arguments notwithstanding, 'such stillness' is not always simply a negative counterpart of 'the mobile'. The 'new mobilities paradigm' strongly emphasises the dynamic relations between mobilities and stillness (Adey 2010; Merriman 2014). Mobility and stillness are socially constructed and mutually constituted. Bissell and Fuller (2011) for example contend that the 'still' can 'do' and influence human practices, and enact particular ways of being in the world. For example, slow-paced tourism and relaxed lifestyles are increasingly popular and considered desirable ways of living (Fullagar *et al.* 2012). In Sanya, many locals celebrate slow-life and maintain a sense of place in spite of Sanya's dramatic socio-economic reconfiguration.

Urban transformation and the changing rhythms of Sanya

Sanya is located at 18°09' N latitude and in the southernmost of Hainan Island in the South China Sea. It covers an area of about 1,918 square kilometres and had a population of about 740,000 in 2014 (Sanya City Bureau of Statistics 2015). However, Sanya was not established as a municipal city in Hainan Province until 1987.[1] In the past, this area was largely ignored by many ancient regimes and remained on the fringe of Chinese empires (Atsushi 1979; Yan 2013). Its vast distance away from the political centres in the Chinese mainland and the geographical barriers of the ocean had not only thwarted the regulations of imperial governments but also impeded inter-regional economic activities and cultural exchange. For a long period, the socio-economic development of Sanya fell behind that of many mainland cities, giving it a socio-political and cultural marginal position in China (Chen and Chen 2016).

In this context, the traditional ways of life in Sanya as well as other cities in Hainan remained relatively 'slow' and closely connected to the location, history and environment of this island. Because of a lack of transport and underdeveloped socio-economic conditions (local residents were mainly employed in farming or fishing), there were few internal and external mobilities of people within the island and between the island and the mainland. Everyday life in Sanya was thus seldom disturbed by people and events 'from outside'. For example, in contrast with the mainland, Hainan Island was historically less influenced by conflict and wars and remained relatively peaceful. Because of the 'static' rhythmic life, it has been argued that local people are not 'so sensitive to time' and lead easygoing daily routines, which were also partly influenced by the yearlong high temperature and lack of obvious seasonal alterations.

The steady rhythms of Sanya have nonetheless been gradually changing since the late 1980s. From 1988, the Hainan authorities obtained enormous support from the central government and implemented a series of policies and reforms to stimulate local socio-economic development (see Brødsgaard 2009).[2] Although these efforts and reforms often led to sharp economic fluctuations in Hainan, they nonetheless brought about a remarkable transformation to the island society. For example, cities like Sanya were able to attract investments and facilitate infrastructure construction although the substantial and constant economic growth often been achieved through tourism. In less than twenty years, Sanya has become the most famous coastal resort and a popular investment place of real estate in China (see Figure 12.1). According to the Report of the Work of Sanya Municipal Government (2016), the GDP of Sanya in 2015 amounted to USD 5.82 billion.

Similar to many Chinese cities undergoing fast economic expansion, urban rhythms of Sanya have been challenged by increasing mobilities of people, capital, information and objects. Each year, thousands of migrants, leisure seekers and business people, among others, flock into this city enabled by improving (inter-) regional transport and tourism mobilities. Constructed in 1994, Sanya Phoenix Airport has become one of the top twenty airports in China, and the largest in Hainan, handling 16.2 million passengers in 2015.[3] In parallel, Sanya Railway Station started to provide a cross-sea train service with the mainland in 2005. Internal island transport has also markedly improved in the past ten years. Hainan eastern expressway was completed in 2007, followed by the western expressway in 2010. At the northern end of the island the circle expressway, and construction of eastern and western high-speed railways were completed in 2010 and 2015, respectively. As a result, mobilities of people and things have been 'speeded up'. For example, it now takes about three hours to travel around the entire island by high-speed train.

FIGURE 12.1 Real estate projects near the beach

Source: Jingfu Chen

In sum, the substantial advancement in transport significantly strengthens (inter-)regional links between Sanya and other places and facilitates population movement, economic activities and socio-cultural interactions.

The polyrhythms of Sanya have also been changed by tourism mobilities and (for some) increasing velocity, seasonality and complexity. According to the Sanya Tourism Development Committee, overnight visitors to Sanya grew from about 3.6 million in 2004 to more than 15 million in 2015, a figure almost twenty times as large as its residential population.[4] The dramatic increase of tourist arrivals and tourism physical infrastructure have raised the velocity of social transformation, economic development, changing urban landscapes and everyday activities of Sanya. For example, Sanya Phoenix Airport has carried out two extended constructions in 2006 and 2010 in order to expand its capability to handle the quick growth of passenger arrivals. Urban demolition, land acquisition and the emergence of new hotels, restaurants and stores has also been seen throughout the city.

In addition, urban rhythms are also characterised by the seasonality of tourism mobilities. During the peak season, large numbers of tourists visit Sanya and bring considerable business opportunities, traffic congestion and other social and environmental problems. This has exerted enormous influence on rhythmic life in the local society. For example, it is common to see tourism employees such as tour guides work overtime in order to make money and local residents suffer from serious inconvenience caused by busy traffic and congestions in downtown areas. In contrast, urban life during the low season appears 'is less vibrant', as illustrated by many empty beaches and the decline in tourism income. The complex urban rhythms in Sanya are thus demonstrated by diverse tourism mobilities, economic activities and many tourism-stimulated socio-cultural practices.

Despite these challenges the dynamic socio-economic reconfiguration and changing urban rhythms are supported by the local authority activities in Sanya. This is underpinned by the dominant discourses of modernity and development in China. In this context, slow transformation and 'slow-life' are often given negative meanings and seen as counter to the goals of advancement and progress. This is aptly illustrated by the official interpretation of the attractive prospect of fast socio-economic development after the launch of the national strategy of International Island Tourism Destination in 2009. This project aimed to stimulate the economic growth of Hainan through tourism development and has directly prompted a new round of investment in Sanya:

> At the end of last year, the State Council implemented the policy, *Suggestions of the Construction and Development of Hainan International Island Tourism Destination*, and provides tremendous support [for Hainan's economic development]. This creates a crucial historic opportunity for Hainan and Sanya. Since 1998, Sanya has achieved double-digit economic growth for twelve years, and continuously improved the quality and quantity of local economy and rural-urban infrastructures. This greatly strengthens our basis of future development. At present, Sanya is under a strategic opportunity. The driving force of socio-economic development in Sanya will be further increased given the

> fast construction of key districts and projects and the rapid improvement of people's wellbeing.
>
> *(Report of the Work of Sanya Municipal Government 2010)*[5]

However, while the economic impact of mobilities on individuals' everyday life is unequal, official positive representation of changing urban rhythms is not always fully accepted by local residents. As Massey (1993) contends (im)mobilities are practised and experienced in different ways as the next section discusses.

'Outdated' lifestyle? Leading traditional ways of life

The diverse urban rhythms of Sanya through mobilities often contrast starkly with traditional ways of living in local communities. For example, many local residents are involved in processes and practices of rapid urbanisation and unlike their ancestors, follow different life paths. For example, many young people leave agricultural sectors to work in urban areas. Others continue to undertake traditional work and leisure routines. For instance, many native residents rely on farming and fishery or run their own small food stores to earn family income. This is particularly evident among rural populations with many middle-aged and elderly urban residents. Other traditional ways of living in local communities such as *Wuxiu* and tea-drinking practice nonetheless remain more broadly popular.

Wuxia comprises an important component of a traditional 'slow-life' in Sanya. Most local residents prefer a snooze after lunch, lasting for one to two hours. People rest at home, in workplaces and even in public areas, taking a break from their activities (see Figure 12.2). It is not uncommon to see workers such as motorcycle

FIGURE 12.2 Two officials having an afternoon nap at the beach

Source: Jingfu Chen

taxi drivers asleep on lawns and benches in public parks. This habit varies across social groups and can be partly attributed to tropical weather and high temperature at midday. The popularity of *Wuxiu* contrasts to 'accelerated' everyday rhythms and the hustle and bustle of urban life in the daytime. Nonetheless, between 12 p.m. and 2 p.m., the city seems to slow down and appears 'inactive' and many commercial activities temporarily cease.

Wuxiu is certainly not unique in Sanya and not everyone in this city takes an afternoon nap. For example, non-local people such as tourists and migrant workers generally spend less time enjoying *Wuxiu*. Moreover, many migrant workers work all of the day in order to increase their income. In comparison, local residents can appear 'casual and easygoing' but in contrast local people are also known for often 'staying up late at night and getting up late in the morning'. Local residents are often criticised for being 'indolent' because they do not make full use of time to make money and work hard. This is illustrated by the opinion of Mr Hu, a mainland migrant, who works as a tour guide in Sanya:

> Local males are lazy! They can go out with friends to have supper in night markets at 1 a.m. and wake up at 10 a.m. How can they get rich in this way? I cannot act like them! During the peak season, I usually get up at 5:30 a.m. and work until 11 p.m. It is difficult to make money.

In addition to *Wuxiu*, the traditional slow-paced lifestyle is also exemplified by tea drinking, which is especially popular among men. It is easy to find local residents gathered at home or in simple tea-houses and engaged in lengthy conversations while drinking tea. One favoured location for tea drinking is Father Tea (*Laoba cha*), which was originally enjoyed by middle-aged and elderly men but is increasingly popular among many women and young people as well.[6] This leisure practice is often conducted in (in)formal and inexpensive public tea-houses, usually located in old urban districts, filled with customers each afternoon – the favourite tea-time in Sanya as well as in other places in Hainan. People dress casually and speak loudly, discussing a wide range of issues including family disputes, lottery results and national and neighbourhood news (see Figure 12.3). Tea-houses can become very crowded and noisy. However, this does not dampen the enthusiasm of tea drinkers, who can stay and hold conversations with friends (and sometimes strangers) for several hours without doing anything else. The casual atmosphere and the popularity of Father Tea was reported by *Hainan Daily* (Sai 2015):

> Tea houses are always bustling and attract a huge crowd. Looking through the cloud of smoke and water vapour, servers busily run back and forth in the hall, filling up teapots for one table and sending snacks to another. They are enthusiastic and considerate. Many customers may only pay USD 44 cents for a pot of tea and stay from the early morning to the dusk. But waiters/waitresses still provide warm and faithful service – nobody will ask them to leave.

FIGURE 12.3 Father Tea

Source: Jingfu Chen

The inclusive and casual environment in tea-houses was also depicted in *Hainan Today Magazine* (Wang 2013: 42):

> Customers of Father Tea can come from any social status. Whether or not they are well-off, all are welcomed to sit down and have a cup of tea. Nobody is excluded. Tea drinkers are free to decide how much they want to order. They can simply pay USD 15 – 29 cents for a seat and do not need to spend money on other foods [such as desserts]. Shopkeepers show a high tolerance of customers' behaviours. Some people only stay in tea-houses for several minutes, while others may stick around for a whole day.

Similar to *Wuxiu*, relaxed tea drinking practices are also considered 'inappropriate' and viewed as a waste of time by many people, especially mainland migrants. They argue that local residents should take advantage of the rapid economic growth of Sanya and be fully focused on improving their living standards. The popularity of Father Tea even appears problematic since the majority of local residents are not 'well-off'. The average yearly income of urban residents in 2014 was about USD 3,909 (Sanya City Bureau of Statistics 2015), which was below the national average. The indulgence by a number of young and middle-aged males in the laid-back leisure of tea drinking is thus believed to unproductive and relaxation in tea-houses is considered 'lazy' and interpreted as a negative attitude towards life.

Such contested meanings attached to the slow-life in Sanya reflect anxiety related to modernisation and development. For most migrants and visitors from the mainland, the common memories of hardship and social changes throughout China during the past decades affirm their belief in 'hard work'.

Conclusion

This chapter has discussed complex and contested urban rhythms through consideration of (im)mobilities in Sanya and explored unequal power relations of modernity in everyday life (Lefebvre 2004). In the case of Sanya, while for many 'speedy' transformation is viewed as active and progressive, the traditional slow-life in local communities, as illustrated by *Wuxiu* and tea-drinking practice, are often depicted as arrhythmic and 'out of step'. In accordance with the discourses of modernity and development, slowness becomes pathological and undesirable. As Cresswell (2006) contends, mobilities and stillness are socio-culturally constructed and acquire different significances with many local practices being stigmatised and local people encouraged to adopt 'proper' and 'more productive' lifestyles. In this sense, the exploration of urban rhythms in Sanya offers important insight into the dominance of urban modernity in contemporary China, which, as Lefebvre (1991a) has pointed out, constantly regulates, colonises and is (re)produced in everyday life.

My research has also contributed important theoretical and empirical insights which add value to critical perspectives on Chinese urbanism by specifically focusing on small cities. Against the backdrop of an expanding national economy, Sanya has quickly grown from a 'marginal' city to occupy an important position in the national imagination in less than twenty years. Indeed, like Sanya, many other small Chinese cities, which were previously not seen as favourable for investment, are now being considered as potential drivers of economic growth. However, by focusing on 'slow-life' in Sanya, I have also highlighted how the rapid development of small cities can nonetheless lead to contested socio-cultural change and the assertion of unequal social relations in small cities. As urban development initiatives move beyond the great metropolises of China, it is vital that more theoretical and empirical research is attuned to urban life in small Chinese cities in order to offer insights that can seek to contribute to more progressive urban change.

Notes

1 Hainan province was established in 1988, the most southern province of the People's Republic of China. It is constituted by Hainan Island and archipelagos in the South China Sea.
2 For the central government, reforms in Hainan have been viewed as a crucial driving force for national economic reform since 1978.
3 Source: the homepage of Sanya Phoenix International Airport (http://www.sanyaairport.com); access date: 20 November 2016.
4 Over 97 per cent of visitors in Sanya are Chinese people from the mainland.
5 Source: The Report of the Work of Sanya Municipal Government (http://www.sanya.gov.cn/publicfiles/business/htmlfiles/mastersite/jhzj_zfgzbg/201307/103845.html); access date: 20 November 2016.
6 Not only in Sanya but also in many other places of Hainan province – Father Tea is very popular.

13

CHINESE URBAN INFORMALITY AND MIGRANT WORKERS' NEGOTIATION OF WORK/LIFE BALANCE

Gengzhi Huang, Tao Lin and Desheng Xue

Introduction

Across the urban world large numbers of people make a living from informal economic activities and house themselves in informal settlements (Davis 2004; UN-Habitat 2009). In China, the size of the urban informal workforce is estimated to be around 200 million people – 50.7 per cent of total urban employment (Huang *et al.* 2016), with a large number of informal workers living in 'urban villages' (Wu *et al.* 2013). The urban informal economy thus clearly plays an important role in cities throughout China (Huang 2009) and understanding the dynamics of urban informality is thus of central importance to developing critical perspectives on Chinese urbanism.

Urban informalities involve economic, political, social, cultural and spatial practices beyond or insufficiently attended to by regulatory systems. This chapter is specifically concerned with economic forms of urban informalities, with a focus on how rural to urban migrants seek to achieve work/life balance. In academic literature, informal work over a long period is more often than not viewed in terms of scarcity of waged employment in the context of 'underdevelopment' and inefficient industrialisation in developing countries. Informality thus acts as a survival strategy or as a buffer against unemployment (Hart 1973). Informal work is also depicted as a 'stepping stone' by which migrants can seek to engage with formal economic systems. In this chapter we nonetheless argue against the idea that informal work is a 'last resort' for migrants looking to survive in the absence of urban jobs. Based on a critical reading of current mainstream perspectives on informal work and empirical research into two forms of informal work (i.e. informal or self-employed entrepreneurs and 'day-workers'), we argue that informality can often be based on a choice of migrant workers' negotiation and balancing of economic and social needs. Specifically, by informalising working practices migrant workers seek, on the one hand, to secure and govern their economic income, and on the other hand, to lead a life with autonomy and flexibility to meet both their family commitments and individual preferences. In doing so we show how informal working practices enable migrants to negotiate the balance of work/life in urban China and call for more research on the articulation of migrants' informal working practices and their everyday life to better understand the meaning of informality in Chinese urbanisation.

Informality, development and urbanisation

Four mainstream perspectives on the informal economy have emerged since the 1970s which provide diverse and sometimes competing understandings of the origins and development of informal economies – dualist, neo-Marxist, legalist and voluntarist (Kucera and Roncolato 2008; Rakowski 1994). In this section we critically review key arguments from each of these schools of thought.

First, the dualist view emerged in the early 1970s when cities in the developing world faced immense poverty and unemployment as a result of the unprecedented population growth, large-scale rural-to-urban migration and insufficient industrialisation (Williams and Round 2010). In many developing countries, a large number of rural migrants moving to cities we depicted as not entering the formal sector, but instead earned income by engaging in various informal sector activities (Hart 1973; ILO 2002). This view outlined an understanding of urbanisation in developing countries which shifted attention from problems of unemployment, over-urbanisation or pseudo-urbanisation in order to consider how urban populations including rural migrants made a living from informal practices in response to a lack of state support. Research highlighted that despite being limited by a lack of linkages with the formal sector, urban informal activities could often be considered as economically efficient and profit-making (Rakowski 1994). Urban informal sector activity was thus foregrounded as a driving force of urbanisation in developing countries by providing income for rural migrants in an urban economy unable to offer them formal employment opportunities. In other words, informality was argued to play a vital role as an employment 'safety net' in urban development. In this literature informal employment was not considered however as a representation of individual initiative, but as a mechanism of the structural forces of unemployment and underdevelopment. Arguments suggested that once sufficient jobs became available in the urban economy, informal workers would automatically take formal employment (Sanyal 1988; Chaudhuri 2000). These dualist perspectives are in essence structuralist.

The neo-Marxist school of thought which emerged in the late 1980s is also structuralist, theorising the informal economy as a by-product of capitalist economic restructuring characterised by a rise in subcontracting economies and flexible arrangements of employment at global and local levels (Sassen 1997). Informalisation in this writing was often seen as a strategy used by capitalists to avoid labour regulation, reduce costs and weaken the power of workers and unions (Castells and Portes 1989; Whitson 2007). For this reason, despite earning an income, workers in the informal sector tended to receive fewer benefits, suffered lower wages, experienced poorer working conditions and had precarious employment (Olmedo and Murray 2002; Siegmann 2016). However, an increasing erosion of job quality and security also ensured that a sizable number of waged workers moved towards alternative informal working practices such as self-employment, in response to a 'predatory' free market (Biles 2008; Portes 1997). In neo-Marist arguments informal workers, especially those who are migrants, are essentially victims of neoliberal capitalist accumulation.

In parallel with such neo-Marxist writing, a legalist perspective was developed by Hernando de Soto (1989). Legality/formality in this work was considered as a privilege of those with economic and political power, while informality or extralegality was the one choice for those who are disadvantaged (de Soto 1989; Maldonado 1995). Advocates of this school agree with dualists and neo-Marxists on the point that informal activities serve as a safety net for social tension and are economic survival strategies for the urban poor. However, legalist perspectives attach importance to the ingenuity of the individual, arguing that the informal economy is underpinned by spontaneous and creative responses to the state's incapacity to satisfy the basic needs of the impoverished masses in the process of urbanisation. In these terms however, the urban poor are shown to contribute significantly to national economic and urban development, with informality thus being seen as a path to development (Bromley 1990).

Since the turn of the twenty-first century, a group of theorists focused on Latin American cities and supported by the World Bank have developed a voluntarist perspective on informality (Biles 2008). This work considers informal employment, particularly self-employment, as a voluntary option of workers as they pursue the advantages of informality – flexibility, autonomy and freedom – and that such characteristics are absent from formal employment (Maloney 2004; Perry and Maloney 2007). In these terms informal workers opt for their occupations according to their individual needs and abilities (Perry and Maloney 2007). For example, female workers forgo formal salaried work in order to balance work and family responsibilities. It is argued that despite low income and little stability associated with the informality (self-employment), informal jobs are a better options than the corresponding jobs in the formal economy (Perry and Maloney 2007). Like legalists, voluntarists attach importance to the initiative of informal workers and their preference for the benefits and nonpecuniary characteristics of jobs (e.g. flexibility, autonomy). This work adds a social dimension to the meaning of informality.

So how can we assess the strengths and weaknesses of these different approaches to understanding the informal economy? While dualist and neo-Marxist perspectives are focused on structural factors (i.e. insufficient formal jobs, restructuring of capitalist economy etc.) driving the expansion of the informal economy, legalists and voluntarists attached importance to workers' individual ability to meet their own work/life balance. However, these perspectives are by no means contradictory. For example, some scholars have pointed out that these seemingly divergent schools are instead complementary because each of them explains diverse causes and effects of a heterogeneous informal economy (Williams and Round 2010; Adom 2014). In this chapter we also draw on a combination of these perspectives to better understand the informal economies of migrant workers in China. In this combination, dualism and neo-Marxism offer important insights into structural forces such as lack of jobs and deterioration of job quality spurred by restructuring capitalism, while legalism and voluntarism highlight labour agency in improving an individual's life.

For example, the significance of informal work in Chinese urbanisation can be primarily, if not solely, recognised as a survival strategy for the large-scale rural-to-urban migration (Huang and Xue 2009). It has been widely argued that a basic conflict of China's urbanisation is the gap between the unlimited supply of rural labour and the limited ability of cities to offer migrant workers employment (Zhou 2010). Research has shown that urban informal employment serves as an antidote to this contradiction to ensure economic and social stability. Indeed, Huang *et al.* (2016) have proved that in China the growth of urban informal employment by 1 per cent contributes to an increase of urbanisation rate by 0.1 per cent. However, this structuralist view offers few insights into non-economic meanings of informal work for rural-to-urban migrants. For rural migrants, moving to cities not only can enable possibilities of better income opportunities, but also brings conflicts regarding family responsibilities. For instance, migrants are often required to return to their hometowns to look after their parents and in some case to help with farm work during the harvest period. Female migrants also take responsibility for childcare/housework. As such in this chapter, we will show how complex work/life balance issues are negotiated by migrant workers in their working practices in seeking to ensure economic security and social mobility.

Following a brief introduction to the methodology and research design that underpins the research presented in this chapter we then draw on a combinational theoretical framework to examine how both structural forces and individual needs affect informal working practices of migrants. More specifically, in subsequent sections we foreground the ways in which informal activities allow migrant workers to negotiate a work/life balance in contemporary Chinese cities (see Chapter 12, this volume).

Methodology and research design

Over the past few decades China has witnessed continuing large-scale rural-to-urban migration, giving rise to a large number of migrant workers in cities (*nongmingong*, 农民工). Guangzhou, located in south China, has been one of main destinations of migrant workers with 6.5 million arriving by 2010–46 per cent of the city's total population (Xue and Huang 2015). Manufacturing sectors that attract migrants include the automobile, electronics, textile and leather industries, as well as construction, retail, catering and hospitality. In this chapter we focus on migrant workers in the leather industry located in Shiling Town, northwest of Guangzhou (see Figure 13.1). The industry has more than 300,000 workers employed in thousands of factories.

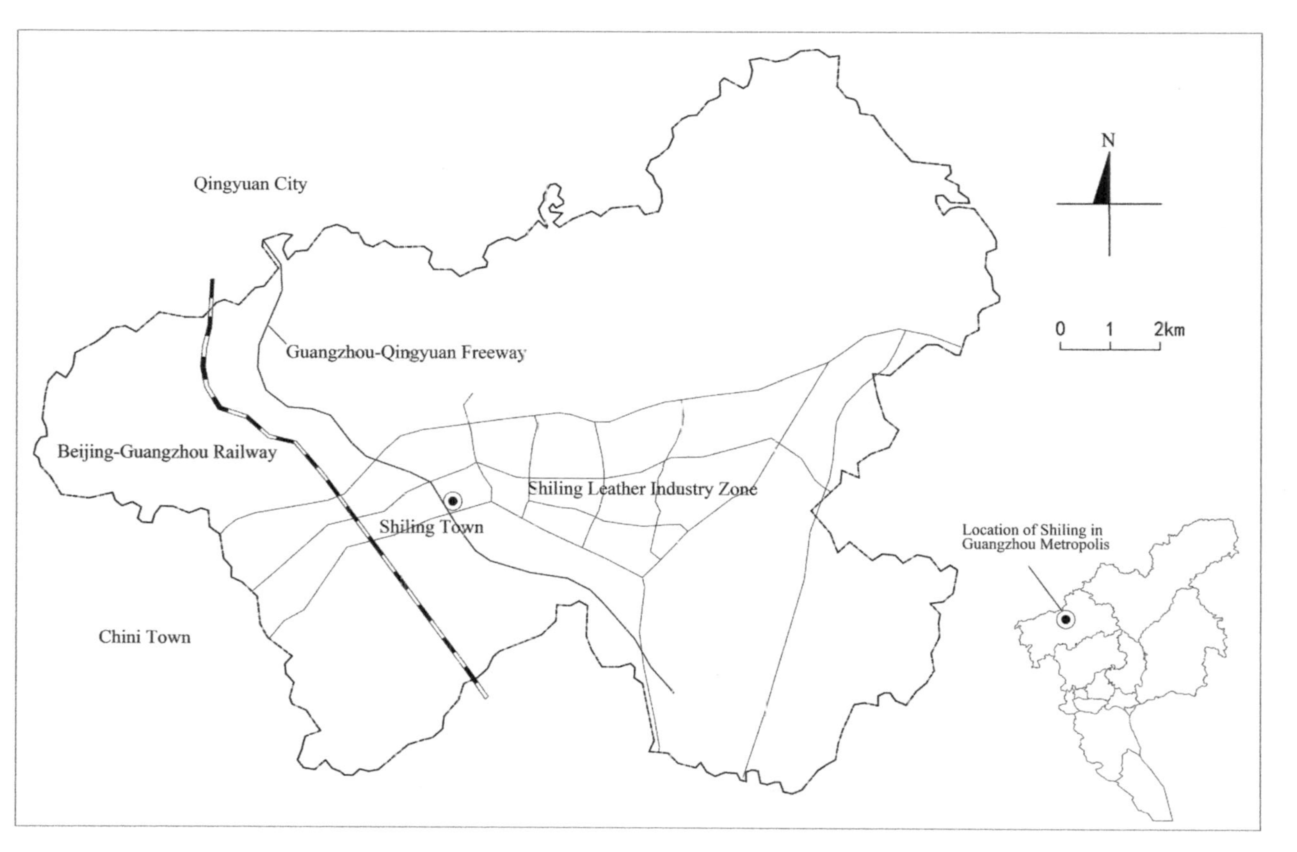

FIGURE 13.1 The location of the leather industry in Guangzhou

Source: Gengzhi Huang

Our research defines informal work as employment where a labour contract is not formally signed and we focus on two types of informal workers: self-employed factory owners without business registration – and with no more than tem workers employed without a labour contract; and 'day-workers' who do not sign a labour contract but get employed on a daily basis. In contrast, the formal economy is defined as economic units (enterprises) with business registration and workers employed with signed labour contracts. These include large and medium enterprises with more than 100 employees, as well as small enterprises with dozens of workers.

Our study included thirty in-depth interviews with informal self-employed factory owners across eighteen different types of processing sectors in the leather industry and thirty-eight interviews with day-workers. Furthermore, we also interviewed six managers in formal enterprises and one officer from the Shiling Investment Service Office. Participant observation was also undertaken in several factories. Respondents' identities have been anonymised.

Informal workers in Global Production Networks

The leather industry in Guangzhou is export oriented, with 70 per cent of leather products exported to more than 130 countries throughout the world. The globalisation of the industry is characterised by the economic linkages between international brand companies, international trading companies, domestic (local) independent brand companies and the global leather market (see Figure 13.2). International brand companies and international trading companies do not directly produce leather. They undertake research, development and sales while subcontracting production tasks to the large and medium production companies in Shiling. The subcontracting system links local production activities with the global leather market.

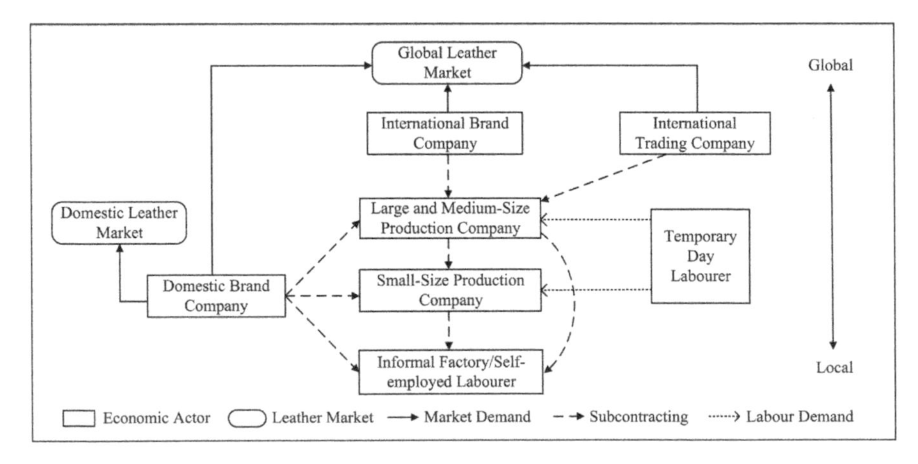

FIGURE 13.2 Informal workers in global leather production networks

Source: Gengzhi Huang

By setting up points of sale or distribution in overseas markets in Latin America, the Middle East and Africa, local brand companies enter the global market. Some local brand companies in relatively large-scale factories undertake all or most production tasks, but some focus on research and development, design, management and sales while outsourcing production tasks. Production firms often do not finish all production tasks, but also subcontract some of them to small production firms and self-employed workers or informal workshops. These small firms also subcontract part of the production tasks they receive to informal workshops. As a result, a hierarchically functional linkage between informal economies and formal economies has emerged. Through these linkages, self-employed factory owners and day-workers are engaged in the division of labour which underpins the global leather industry.

According to official estimates, there are at least 1500 self-employed workers or informal workshops/factories in Shiling. These informal workshops have no business licence, no tax registration and no registered brand. They often specialise in one or two segments of the production chain, such as metal processing and electronic embroidery and can be characterised by three or four workers employed in a small workshop (see Figure 13.3). These informal workshops are often family owned and about a quarter of them hire employees based on kin and friendship networks.

FIGURE 13.3 A typical informal factory in Shiling, Guangzhou

Source: Gengzhi Huang

Official estimates also point to around 1,000 (mostly male) day-workers who respond to seasonal variations in the global leather market. Each morning day-workers congregate at a road junction and when 'recruiters' arrive the day-workers negotiate details of the nature of the work and payment. Such an arrangement for day-work is conducted in an unwritten way – without a formal contract. After completing the work task, workers get the payment based on 'piece-work', leave the factory and wait for the employment of the next day.

Economic and social motivations of informal workers

Acting against poor working conditions

As production firms in Shiling are mostly at the low end of the global commodity value chain, low wages are a main strategy to enable a competitive advantage in the global market. As a result, a large number of non-skilled workers earn a meagre salary of between 1500 – 2000 yuan per month, working twelve hours each day and with only one or two days off each month. In order to prevent workers from leaving, firms hold a month's salary as security that workers lose if they leave without agreement. Moreover, employers in small firms often delay wage payment due to problems of capital turnover, especially when production orders decrease. These working conditions are not specific to Guangzhou – a recent research report on Chinese migrant workers at the national level showed that 46 per cent worked nine hours a day or longer; 52.2 per cent experienced wage delays; and 76 per cent did not get any overtime pay (Wong 2011).

In facing such capitalist exploitation some workers choose to protest, but generally receive little significant political support due to the absence of effective trade unions and the state's privileging of capital for the sake of economic development (Pun and Lu 2010; Chan 2014). Some workers move to new employers despite the likelihood of similar working conditions and rates of pay. However, in addition to these responses, many workers choose to exit from regular waged work and turn to informal activities such as self-employment and day-work. Our research shows that most self-employed workers are motivated by the desire to avoid the fate of cheap labour. They complained that the wage in factories are too low and work hours are too long. This motivation was also stronger within the group of day-worker who had exited waged employment. As one day-worker suggested:

> I often got no payment for one or two months and even for several months. I didn't get paid for two months in the last factory I worked for. I just saw the boss in the morning, and he disappeared suddenly in the evening. My and my wife's salaries, which amount to 10,000 yuan together, were lost. The factory equipment is not valuable, and is worth only several thousand yuan. So I dare not enter the factory any more.

Most of factory workers we interviewed had lost their deposit held by their employer, often being subject to violent response by security guards at the factories.

By contrast, day-work seemed to have higher and more secure incomes. According to one day-worker, they could get 120 yuan by working eight hours, while a regular waged worker might need to work more than twelve hours to get the same; moreover, workers often suggested sentiments such as:

> I'm not afraid that the boss will not pay. If he doesn't pay, I just lose one day's salary. Moreover, we day workers here got united. Last time one boss didn't pay the salary, when he came here to recruit workers the next day, we turned his car over. So, generally, bosses dare not withhold our salaries.

As such, engaging in informal work was not the last resort of workers responding to a lack of jobs, but allowed them to escape waged employment as neo-Marxists have argued. The self-employed workers sought to break away from the exploitative employment relationship, while day-workers gained power to negotiate with employers. However, it should be noted that the self-employed and day-workers were still strongly dependent on the performance of the leather economy as a whole. Indeed, given the linkages between formal and informal economies, the self-employed and day-workers are often the first to be affected when orders decrease.

Balancing work and family commitments

In addition to avoiding the pitfalls that characterise the working conditions of cheap labour, most migrant workers engaged in self-employment in order to attain autonomy and regain control over their lives and to respond to family commitments. As one self-employed worker who runs a small sewing workshop suggested:

> If I work in a factory, nobody looks after my children. If I do not work at all, the salary earned by my husband can't cover our living expenses. So I opened the shop with others to make some money, so that I can make a supplement to my household income while being able to take care of my children and husband.

Informal work is thus an effective way for migrant families to gain some economic security as well as allowing time for household duties (including taking care of children, cooking dinner, cleaning, and etc. – see Figure 13.4).

Such informal employment also allowed workers to escape the rigid organisation of factories. As one day-worker in his twenties suggested:

> The factory management is too strict. You often have to work overtime, though you are so tired. The boss often yells at you, and the salary is low with only 60 – 70 yuan per day. By contrast, we now can earn no less than 90 – 100 yuan per day. So I could just work for 20 days per month if I like, and I can have a rest whenever I don't want to work. I feel much freer than in the factory.

FIGURE 13.4 Members of a knitting group

Source: Gengzhi Huang

Preference for a more 'free life' was also specifically embraced by the so-called 'new generation migrants' who were born after 1980. Compared to their predecessors born between the 1950s and the 1970s, the new-generation migrant workers often have a higher expectation of life/work balance, and are more active in opposing unfair treatment (Pun and Lu 2010). These migrants know little about traditions of farming, have received higher levels of education than their parents and are strongly attracted by urban life associated with consumption, shopping online and ChinaNet (see Chapters 6 and 8, this volume). This new wave of migrants are more likely to undertake informal work as an alternative to precarious waged work and in order to ensure economic independence and to maintain some sense of autonomy.

Conclusion

This chapter has explored the motivations of migrant workers participating in informal economies in the Chinese city of Guangzhou. Our research has highlighted that participation in informal economies is not only a result of workers' responses to unemployment and poor working conditions, but is motivated by their pursuit of a more flexible and 'free life' (see Chapter 12, this volume). However, we have also highlighted the heterogeneous nature of migrants' experiences with regard to gendered and generational dimensions. In terms of gender, informal work enables female migrants to negotiate work and family commitments. Generational differences highlights 'new-generation migrants' who choose informal opportunities against waged formal work which is more rigid, precarious and often poorly remunerated. Informal work thus meets their individual preference for 'freedom and flexibility' in comparison to regular waged work.

Finally, despite the positive opportunities and motivations for migrants to choose informal work, it should also be noted that informal workers nonetheless suffer low income and lack of social protection, and moreover, are vulnerable to changing economic conditions at global and local levels. More research needs to be done to better understand how informal workers are linked to the global and local economic systems. How do migrants cope with livelihood crisis in periods of economic recession? What is the role of (in)formal work across the life course? These and many more questions need to be addressed in order to advance critical perspectives on the role of informal economies in Chinese urbanisation. To that end, this chapter highlights the need for more nuanced critical theoretical perspectives complemented by empirical research that focuses on the heterogeneous nature of urban informal economies unfolding in contemporary Chinese cities.

Acknowledgements

This research is supported by The High-level Leading Talent Introduction Program of GDAS(2016GDASRC-0101), The Scientific Platform and Innovation Capability Construction Program of GDAS (2016GDASPT-0210), the Science Foundation of Guangdong Province (2015A030313842) and the National Science Foundation of China (41401169; 41320104001).

PART V

Bodies, emotions and atmospheres

14

EMBODYING CHINESE URBANISM

Mark Jayne and Ho Hon Leung

Introduction

Rapid urbanisation in China over the past decade has taken place in parallel with increasing academic interest around the world (Wu 2009; Whitehead and Gu 2006; Lu 2011; Weiping and Gaubatz 2013). While theoretical and empirical writing about Chinese cities mirrors complex and diverse international urban research agendas, here we point to the importance of critical understanding of urban life (see Jayne and Ward 2017). As this volume shows Marxist, feminist, post-structural, post-colonial and non-representational thinking is yet to significantly impact research focused on Chinese cities. Advancing this project we consider everyday social and cultural practices and processes. De Certeau (1984) and Lefebvre (1971) highlight the significance of the city as a site of struggle between the powerful and less powerful and that

> the point of departure for critical social theory should always be everyday life, the banal, the ordinary … changing everyday life: this is the real revolution … and any point has the potential to become central and be transformed into a place of encounter, difference and innovation.
>
> *(Schmit 2012: 58)*

Despite increasing academic engagement, social and cultural theory is marginal compared to political, economic and spatial analysis of Chinese cities (although see Kenworthy-Teather 2001; Kong 2011; Crang and Zhang 2012; He 2013; Wang *et al.* 2013). As noted in the introduction to this book, it is also important to acknowledge significant writing by Chinese scholars, not published in English language journals and books, offering rich and detailed understanding of social and cultural life in Chinese cities including a diverse range of topics, case-studies and contexts. However, as Pow (2011: 47) argues that such writing tends to fall into a trap of assuming political, economic, social, cultural and spatial urban practices and processes 'peculiar to China have rendered the Chinese urbanization trajectory *more different* than similar from Anglo-American cities ... [and accepted rather than critiqued that the] Chinese state in particular is seen to respond to and/or create conditions and institutions that render urban China's experience as unique and exceptional'. Such comments notwithstanding Ward (2009) points to progress that can be achieved by 'theorizing back' to work on Chinese cities with reference to critical theory. This offers opportunities to reflect on the geographically uneven foundations of contemporary urban scholarship. Moreover, this ontological and epistemological agenda is vital to exploring how historical and contemporary Chinese scholarship can contribute to understanding urban spatiality as distinctively situated social relations, practices and institutions whilst recognising that urban life is bound up with global processes. Indeed, Edensor and Jayne (2011) suggest there is much to be gained from theorising urbanism as mobile, fluid and relational, and territorial and fixed, being continuously contested, perceived and conceived according to a host of urban imaginaries around the world.

Here we signpost how a research agenda focused on theorisation of bodies and embodiment can contribute to understanding Chinese urbanism. The chapter begins with a review of debates relating to embodied geographies. We then draw on this writing to discuss public dance and massage with reference to broader theoretical work relating to public/private space, individual/collective practices and experiences and comfort/discomfort. In conclusion we reflect on how critical perspectives on embodied geographies of urban China can contribute to research within and beyond urban geography.

Embodied urbanism

Studies of bodies and embodiment have contributed to the advancement of non-(and more-than) representational geographies (Thrift 2007; Anderson and Harrison 2010). Johnston (2009: 326) for example suggests:

> the body is a crucial site of sociospatial relations, representations and identities. It is the place, location, or site of the individual. It is also a site of pain, pleasure, and how other emotions such as well-ness, illness, happiness, and health are constructed. The body is the location of social identities and differences such as gender, sexuality, race, ethnicity, age (dis)ability, size shape, appearance and so on. These identities or subjectivities may form the basis for oppression and exclusion, meaning that the body as a space is bound up in knowledge and power. The body is, therefore, a site of struggle and contestation. Where a body can go, what it can do, who can get access to it, is regulated and controlled.

Research has considered fluidity of bodily boundaries in relation to 'ugly' and abject bodies, pregnant bodies, chronic illness, old age and so on (Kenworthy-Teather 1999). Key arguments suggest emotions reside in bodies and places and exist as relational flows, fluxes and currents, in-between people and places. Theorists suggest bodily boundaries are frequently transgressed in emotionally powerful, disruptive and conflictual ways (Longhurst 2001).

Non-representational work on emotional and affective geographies is closely aligned with thinking on embodiment. Theorists have considered 'embodied and mindful phenomena that partially shape, and are shaped by our interaction with people, places and politics' (Davidson and Bondi 2004: 373). Research into joy, sadness, fear, love, hate etc. has highlighted how emotions matter to geographical research. Davidson *et al.* (2005: 10) suggest our lives can be bright, dull or darkened by our emotional outlook and unpacking emotions (experientially and conceptually) must be understood with reference to 'socio-spatial mediation and articulation rather than as entirely interiorised subjective mental states'. Geographical work on affect has similarly avoided focusing on 'interiorised subjective mental states', instead attending to human interaction with 'things' that make up the world – assemblages of non-human actors, technology, the built environment and so on (Anderson and Harrison 2010).

Whilst there is overlap between geographies of embodiment, emotion and affect there has been debate about their limitations. Thien (2005) suggests that affect thinking about human/non-human interaction has marginalised geographies of the personal, masking emotional subjectivity and eclipsing corporality, intersubjectivity and politics of position. Tolia-Kelly (2006) and Hemmings (2005) argue that cut loose from life worlds affect can be anchorless and autonomous and difference in subject position is erased with histories of powerlessness and subjection forgotten. In describing differences between embodied/emotional and affectual geographies, Pile (2010) summarises this point of departure suggesting the former seeks to consider a politics and ethics of 'caring or of emotional transformation … [while the latter focuses on] the manipulation of affect by powerful elites, covertly behind our backs' (Pile 2010: 5).

Writers such as Cresswell (2006b), Lea (2007), Lim (2007), Jayne *et al.* (2010) and Edensor (2012) have productively worked at the intersection of these approaches. This work shows how registers of expression, immanent forms of connection and association offer the promise of liberating ethics, a progressive politics of social difference that can creatively change social relations. Such writing advances understanding of fluidity of boundaries between emotions, embodiment and affect, what Venn (2010) calls '"mind-body-world" where human existence is characterised by instantaneous correlation of "information": facts, signals, rumours, news, mixed in with moods and emotional energies, enabling agents to participate in an activity in which they behave both as an individual and as an element of a collectivity' (Venn 2010: 132).

In the remainder of this chapter we highlight how research focused on embodiment, emotions and affect offers fruitful avenues for understanding political, economic, social, cultural and spatial practices and processes which constitute Chinese urbanism. By sketching theoretical insights into embodied (and emotional and affective) geographies of dance and massage this chapter contributes towards reconciling abstract and subject-less approaches to embodiment with subject-sensitive research. We draw on this critical tension to discuss dance and massage with reference to broader theoretical debates regarding public/private space, individual/collective practices and experiences and comfort/discomfort. In doing so we show that unpacking the 'mind-body-world' (Venn 2010) with reference to assemblages of (non)human actors can offer valuable insights the complexities of urban China.

Embodying Chinese urbanism

Over the past two decades Chinese cities have experienced remarkable change. Construction of new cities, physical redevelopment including mixing of new and historic urban spaces and places has taken place alongside changing social structures, relations and everyday life. When combined with economic growth, increasing population density and continued dominance of one-party political rule it is clear that urban researchers must take up the challenge of engaging with diverse and complex theoretical analysis necessary to better understand changing Chinese urbanism and similarities, differences, connectivities, relationalities and mobilities with cities elsewhere in the world .The reminder of the chapter contributes to that project.

Dance

Towards the end of *The Chinese City* (2012) is a two-page boxed case-study entitled 'Dancing in the Streets'. In parks, underpasses, street corners and car parks, in fact any open public space, Wu and Gaubatz describe how a variety of dance styles are performed by a diverse range of citizens, estimating that there are at least 500 non-commercial public dancing sites in Beijing alone. Chinese *yangge* folk dance, Argentine tango and ballroom are examples of this important recreational activity undertaken in informal clubs and commercial classes. Wu and Gaubatz suggest that ballroom dancing became popular in China during the 1950s – 60s, as an expression of Chinese modernity, a legacy of the Treaty Port era, but was banned during the Cultural Revolution. From 1980 there was a popular resurgence in urban public dancing that continues today. Other writers highlight how ethnic groups in China (Pei 2003; Yang 2006) and Chinese diaspora in Japan (Ferrer 2006) adopt dance as a marker of identity in public space as a prominent feature of urban life. While this writing offers interesting insights, there has nonetheless been an under-theorisation of dance and Chinese urban life.

In contrast geographers have offered insights into symbolic, performative and more-than-representational elements of dance (Thrift 1997; Nash 2000; Revill 2004; Cresswell 2006a; McCormack 2008a). Dance is understood to be rhythmically and stylistically diverse, related to community, ritual, religion, leisure and commercial culture. Thrift (1997: 147) argues that dance is an alternative way of 'being in the world', an articulation of 'complexes of thought, with, feeling that *words cannot name, let alone set forth* ... a way of accessing the world in which the body-subject can symbolize through semblances'.

The empirical insights of Wu and Gaubatz (2014), Pei (2003), Yang (2003b), Ferrer (2006) alongside theoretically focused geographical writing focused on embodiment highlight fruitful avenues towards understanding a key element of dance in Chinese cities – performance and experience of 'dancing in the streets'. Bromley's (2004) work is useful for explaining urban dancing through his assertion that 'private' actions in 'public' domains are often associated with 'informal' practices. Bromley (2004: 294) considers such boundary crossing as a 'moral' logic rather than

> taking possession of public space in a selfish way … [where] popular meanings can be produced through dialogical encounters … respondents look to the material form of the site, and its location, in order to discern the intent of the space and thus shape a moral and aesthetic response to it.
>
> *(Bromley 2004: 294)*

The genealogy of urban dance in China – as an assertion of modernity, its proliferation into public space from the 1980s as a response to the end of the outlawing of the popular pastime – thus can be understood as practice that is 'fugitive, fleeting and value laden' but also more-than-representational in terms of the political and social quality of dance understood with reference to its cultural history and hierarchy (Revill 2004: 206).

While there are diverse Chinese urban dance styles, locations and participants from different socio-economic groups; for example, young people celebrating national Children's Day, and commercial venues' employees 'dancing-as-advertisement', undoubtedly the largest and most visible group involved in dancing is older people. Older people's dancing often takes place in the same spaces as Tai Chi, or vigorous exercise using state-provided equipment and emerged through government sponsored programmes (see Figures 14.1–14.5). Older people's involvement in dance can be considered as an echo of historic government manipulation of bodies, emotions and affect through the kind of organised dance routines used to ensure productive workers, good citizens and to mark national identity, that took place in nations across the world, but particularly in Socialist and Communist countries. However, the role of older people as visible urban actors also has more to tell us about a range of changing urban political, economic, social, cultural and spatial practices and processes in China (see Chapters 2, 4, 7, 9 and 12, this volume).

For example, spaces and places where older people dance relates to historic notions of 'nature' and 'home' (Zhang 2010). Dancing in parks and tree-lined squares and less-formal 'green spaces', such as entrances to underground train stations and roadside residences with discursive notions of 'good life' which historically included physical and spiritual connections and proximity between domestic life and 'nature', which for many is increasingly seen as unattainable in the high-rise physical development that characterises many contemporary cities in China (see Chapter 4, this volume). Miles (2007) also points to re-articulations of Confucianism founded on social order and relationships as virtues of humanity. Miles argues that the emergence and intensification of consumer culture in China since the 1980s is an important element of national identity, economic prosperity and the emergence of China as a world 'superpower'. However, Miles (2007) highlights that proliferation of consumer culture has also been met by concerns over increasing 'individualism' and proliferation of images, symbolism and spaces of 'western' globalising consumer culture. Despite such discourses Miles (2007) suggests older people are beginning to embrace consumption opportunities and urban dance is an example of individuality *and* collectivism underpinning a new 'sense of freedom'. This resonates with Jayne and Ferenčuhová's (2014) research in a (post)socialist Slovakian housing estate, where wearing of 'comfortable' clothing – tracksuits,

FIGURE 14.1 Young people celebrating National Children's Day

Source: Mark Jayne

FIGURE 14.2 'Dancing-as-advertisement' by commercial massage staff

Source: Mark Jayne

FIGURE 14.3 Older people exercising in urban parks

Source: Mark Jayne

FIGURE 14.4 Older women dancing in a public shopping street

Source: Mark Jayne

FIGURE 14.5 Older people night-time ballroom dancing in urban parks

Source: Mark Jayne

nightwear and other garments usually associated with domestic spaces – by diverse social groups in public spaces is popular. Life in state-planned modernist housing led to wearing comfortable clothing in public spaces in order to make them more 'homely' which Jayne and Ferenčuhová (2014) consider an everyday political response to state socialism and later the emergence of consumer capitalism.

Stevens's (2007) work is also useful for understanding collective/individual dancing in urban spaces in China, by outlining everyday political, critical and ethical potential of 'playful' or enjoyable adult practices (see Chapter 11, this volume). Stevens argues that in comparison to 'other' spatial and representational practices play is a response to regulation of public space which suggests a transformative capacity and 'politics of playing that are primarily bound up in experiencing vitality rather than strategic oppositional endeavour' (Stevens 2007: 318). Such theoretical arguments point to complex relationships between individual/collective symbolism, performance and experience of Chinese urban dance. Indeed, Thrift suggests dancing creates emotional and affective urban spaces where 'the subject is also engaged in embodied affective dialogical practices that it is born into and out of joint action' (Thrift 1997: 125). Thrift argues 'dance creates networks' which can elude power – it cannot be commanded, since it is not made up of fixed means – end relationships' (Thrift 1997: 149). This highlights how the spectacle of older people – friends, acquaintances and strangers coming together in the city to dance – is underpinned by unfolding, longstanding and fleeting moments of social interaction. With loud music played often using crackling inefficient public address systems or from portable stereos, framed through bright artificial lights on hot and humid summer nights, older people dancing in Chinese cities signposts how 'the body moves promiscuously with all manner of assemblages, it is a teeming mass of multiplicities' (Deleuze and Guattari 1987).

The popularity of ballroom dancing and Tango in China can also be understood with reference to geographical writing about 'touch' and comfort/discomfort. Social dancing involves sensory, physical and social experiences between bodies and the city itself. Cant (2012: 221) suggests 'embrace' engages social and bodily inequalities through a reciprocal

> metaphysics of intimacy and distance, distinction and separation, of different types of open and closed embraces … [which] take place across generation divides and differences of ethnicity and class, short dancers with tall, large dancers with small and that there is also the intention of dancing with a range of different peoples over the course of the evening.

Urban dance thus offers a world of sensations, of movement, of the loss and recovery of physical control, and a 'collective body-for-fun' (Stevens 2007: 319). In these terms dancing 'is not merely the private, subjective enjoyment of the body, but also a symbolic transformation of feeling with the body … That is to say, they [dancers] inhabit an imaginary world of their own making, central to which is their comportment and that of their fellows' (Radley 1995: 11).

However, bound up with urban assemblages of dance is embodied and emotional comfort and discomfort. Revill (2004: 206) argues that 'dancing profoundly socializes the body, from the embarrassment of excess perspiration to the blur of self-contained exhibitionism as couples pirouette at speed … the dancing body communicates and speaks volumes in response to others'. Similarly, Cresswell (2006b: 55) suggests that dance 'embodies a complex process of exclusion and othering, an account of *correct* movement' that Nash (2000) suggests can underpin social and political inequalities. There is clearly empirical research that needs to be done to understand the complexity of social relations and inequalities in relation to personal and collective participation in Chinese urban dance. Such research is vital in order to understand the ways in which dance enables the performance and experience of 'surprise of the unexpected encounter, the productivity of the encounter that welcomes the future openly' (Grosch 1999: 25) or manipulation of bodies, emotions and affect that reinforces social inequality or to ensure social control.

This section has sketched theorisations of Chinese urban dance. In doing so our engagement with political, economic, social, cultural and spatial assemblages of human and non-human actors have uncovered references to nature, Confucianism and consumer culture that point to how urban geographies in China offer new avenues to nuance, challenge or advance urban theory. Focusing on symbolic, performative, experiential and more-than-representational elements of dance taking place in some of the most densely populated urban areas in the world signposts how critical theoretical and empirical work focused on bodied and embodiment in Chinese cities has much to contribute to urban theory.

Massage

Geographers have recently concerned themselves with spatialities of touch (Paterson 2005; Dixon and Straughan 2010; Lea 2012). Patterson *et al.* (2012: 8) argue that touch has been superfluous or inconsequential in social science scholarship but attention has gathered momentum through work on bodies, emotions and affect. Theorists have thus begun to offer insights into the ways in which 'places of touch are inevitably and sometimes powerfully experientially differentiated' arguing that 'touch is relational, co-produced, is co-constituted in a series of configurations between human and (non) human, and people and spaces alike' (Paterson *et al.* 2012: 8). To a large degree these arguments complement those relating to dance which suggest bodies are not unitary and bounded but are individually and collectively emergent through relations, including the senses. Just like spaces of urban dance, massage venues have become archetypal consumption spaces in Chinese urban shopping and commercial districts. Figure 14.6 shows examples of venues with a diverse offering of massages techniques, design and cost. Bringing together theoretical perspectives on massage and Chinese urbanism thus offers insights into logics of scale and location which 'promise to disclose new topographies of circulation, connection and mobility' (Lorimer 2008: 88). To that end, we discuss dimensions of public/private space, individual/collective performances and experiences and comfort/discomfort in order to discuss embodiment/emotions/affect relating to distance/proximity, pleasure/pain and mobilities/stillness (see Chapter 2, this volume).

Lea's (2012) discussion of therapeutic massage draws on Wolkowitz's (2002: 497) depictions of 'intimate, messy contact with the (frequently supine or naked) body, its orifices or products through touch or close proximity'. Touch is thus 'assumed to be a constant not only within one form of body work (e.g. massage) but also between different forms of body work (e.g. massage and nursing)' (Lea 2012: 300). However, in seeking to 'advance understanding of these experiential, felt, fleshy registers of the body' Lea (2012: 30) points to diversity of massage and differences in touch and technique. Indeed, such recognition is important in interrogating how mind-body-world relationships are 'made' through the action of touch and how the body is made and remade through boundaries of individual and collective bodies. For example, Lea (2012) suggests that there must be a balance between either too much pleasure or too much pain, in order for massage to release 'knots' which develop in response to negative effects of bad posture, repetitive movements and the stresses and strains of everyday life.

In densely populated Chinese cities, with people living, working and playing in close proximity, unintended and regular jostling, bumping, and physical closeness ensures that massage is popular because the 'disordering effect of touch re-arranges the bodily archive such that past experiences might be "released" from their location and eventually released from the body' (Lea 2012: 38). Massage also allows corporal experiences of 'waiting' during 'journeying' through the city. Relationalities of bodies between activity and inactivity can be explained by Bissell's (2007: 277) suggestion that 'the event of waiting should no longer be conceptualized as a dead period of stasis or stilling, or even a slower urban rhythm … but as a variegated affective complex where experience folds through and emerges from a multitude of different places'. This offers insights into formation of subjectivities relating to massage in cities as 'not

FIGURE 14.6 Examples of massage venues in Chinese cities

Source: Mark Jayne

as slowed rhythms that are somehow opposed to speed, but instead are themselves examples of rich durations, that might be banal and prosaic hiatus through virtuality ... that fold through multiple temporalities, allowing us to consider not only their radical relatiationality but also their irreducibility' (Bissell 2007: 279). Visiting massage venues in order to ease the 'hustle-and-bustle' of city life enables a 'corporeal attentiveness to the immediate environment ... the act of bodily stillness through waiting is instrumental in heightening an auto-reflexive *self*-awareness of attention to the physicality of perception of the body itself' (Bissell 2007: 286).

In these terms, massage is important for people to release 'the knots' of city life built up through 'collective' touch of human and non-human actors. Such insights suggest a connection between the body, flesh, skin, emotions, urbanism and experiential and felt qualities of 'therapeutic' forms of touch through massage. Indeed, Lea (2012: 30) argues that researchers must pay attention to 'the idea of intimacy, so that intimate touch means not only touching the "outside" of the body in areas culturally or socially understood as intimate (e.g. the stomach, the face, the ears) but also in affecting intimate and deeply held sensations, feelings and experiences'. Indeed, Paterson (2005: 161) suggests that while 'independence, bodily integrity and self-sufficiency are encouraged in our culture [and we assume he means 'Western'], we also value a more personal, intimate, emotional care in which touch is crucial yet sharply differentiated' which is appropriate in some spatial contexts and with some body parts, but decidedly inappropriate to others. Massage thus offers understanding of intersections of practice, embodiment and performative-based practices of sociality underpinned by Thien's (2005) 'contours of intimacy'. Massage allows suspension of social norms of 'closeness' and interaction. In contrast to other spaces and places in the city, encounters between customer and practitioner highlight how distance/proximity relates to 'the body's capacity to act or be affected', in order to re-invigorate the 'vitality, or potential' (Massumi 2002: 35) not only of people's bodies and emotions but of urban life itself (see Chapter 12, this volume).

While touching human and non-human actors is a feature of everyday urban life throughout the world, the popularity of massage in China ensures that this topic deserves further theoretical and empirical investigation. Indeed, empirical research is needed to unpack theoretical insights and arguments relating to a diverse range of topics and issues, such as the complex construction of embodiment, emotions and affect relating to the role massage in people's everyday lives with reference to class, age, gender, ethnicity and sexuality of both customers and practitioners. For example, one interesting element of Chinese urban massage is that it is not always an individual activity. Commercial massage venues generally have treatment rooms of various sizes where groups of friends, family, colleagues can experience massage when 'doing business', socialising, watching television, drinking tea and enjoying snacks. Consumption practices before, during or after physical therapy of massage offer insights into Chinese massage venues as 'affective space of touch as therapeutic ... empathetic and transformative ... [which] integrate the somatic and sensory experiences of these tactile, therapeutic practices' (Lea 2012: 165).

Sustained theoretical work and empirical research is needed to better understand the ways in which massage allows people in Chinese cities to 'self-medicate' to overcome the stresses if urban living, or seek to momentarily 'escape' tiredness, people, things and atmospheres in the city, individually or with others, to seek out time for relaxation and comfort as a direct consequence of mundane practices of bodies moving through the city (Jayne *et al.* 2012). Empirical research is vital to investigate the popularity of massage in China as a response to emergent urbanism founded on manipulation of embodied, emotional and affective geographies related to rapid urban population growth, physical redevelopment and new globalising economic and cultural accumulation strategies *and* relates to individual and collective caring and emotional transformations. Such a research agenda is important in highlighting geographies of intimacy extending beyond studies of the home and family, or gender, sexuality and sex to consider a broader range of interpersonal encounters that are physically and emotionally proximate in the city (Valentine 2008).

Theorising massage through embodied, emotional and affective urbanism allows understanding of public/private space, individual/collective experience and comfort/discomfort with reference to distance/proximity, pleasure/pain and mobilities/stillness. This argument highlights that massage is important for thinking through the ways in which Chinese urbanism is both a product of elite intervention and is replete with expression, immanent forms of connection and association offering promise of liberating ethics, progressive politics of social difference that can creatively change social relations. Indeed, such theoretical and empirical work is vital in advancing understanding of the similarities, differences, connectivities, relationalities and mobilites between Chinese cities and urbanism around the world.

Conclusion

In sketching a research agenda relating to embodied urbanism in China we are aware that we have inevitably opened up more questions than we have been able to answer. Such comments notwithstanding we argue that theoretically informed empirical research which utilises long-established and new performative methods (Dewsbury 2000; Harrison 2000; McCormack 2008a) is nonetheless vital to fully explore the critical potential of researching Chinese urbanism. Chinese cities are an important resource to chart theoretical and empirical lines between understanding how bodies, emotions and affect generate politics and ethics of caring transformation and/or are manipulated of by powerful elites. The diversity of Chinese cities and changing economic, social, cultural and spatial relations offer fruitful avenues that are deserving of sustained research. To that end we argue that embodied, emotional and affective geographies highlight how theoretical and empirical engagement with Chinese urbanism offers opportunities to nuance, challenge or advance urban geographies that are too often dominated by ideas and case-studies developed with reference to European and North America.

Notes

1 This chapter draws on material published elsewhere: Jayne, M. and Leung, H. H. (2015) 'Embodying Chinese urbanism: towards a research agenda', *Area*, 46 (3), 256 – 267.

15

NOISY CITIES

Jie Zhang

Introduction

Urban soundscapes and the spatial-temporal nature of 'noise' in cities is increasingly becoming an issue of academic, policy and popular concern (see Gallagher and Prior 2014; Gandy and Nilson 2014; Revill 2016). For example, in *The Epic of Gilgamesh* people in the city of Shurrupak were punished by the gods for making too much noise (also see Goldsmith 2012: 21). Philosophers, writers and artists such as Schopenhauer, Dickens, and Lessing have also given 'noise' thoughtful consideration. Similarly, academic research across the social and medical sciences has considered soundscapes with regard to a diverse range to topics including; industrialisation and modernity (Corbin 1998; Thompson 2002; Yablon 2007; Payer 2007); class and resistance to hegemonic power (Attali 1985; Bailey 1996; Picker 2000); personal privacy and social communication (Bijsterveld 2008); mobilities and advertising (Russo 2009); and the spatial-temporal discursive construction of noise as a 'problem' (Keizer 2010; Goldsmith 2012).

In China, there has similarly been long-standing debate about 'noise'. In *Shuo Wen Jie Zi* published during the Western Han Dynasty (one of the earliest dictionaries in the world), *clamour* means *interruption*, that is, a disturbing sound is called 'noise' – or to put it another way 'unwanted sound', 'unmusical sound' or 'everything [the ear] hears that hinders communication' (Richards 1961: 274). More recently, the famous acoustician Ma Da-you suggested that following political and economic stagnation relating to the Japanese and Chinese Civil War's, the 'noise' of economic and cultural activity became an important symbol of recovery after the foundation of People's Republic of China (Ma 1987: 81). Indeed the *Institute of Acoustics*, part of the *Chinese Academy of Sciences* where Ma worked, has undertaken sustained empirical research since the late 1950s, in large cities such as Beijing and Chongqing (which was halted during the Cultural Revolution) and elsewhere in urban China since the 1980s and discussed 'noise' with regard to political, economic, social and cultural and physical change in cities. In this chapter, I advance such critical perspectives on soundscapes by focusing on neighbours, household appliances and traffic to highlight the paradox of 'noisy' city life with regard to embodied, emotional and affective urbanism.

Noisy neighbours

In China, there has been a growing number of media reports of increasing levels of violence (and even murder) between neighbours arguing about 'noise'. Lu Xun, an author writing in the early 1900s who depicted everyday life in Beijing has similarly discussed various encounters in 'noisy' and thus 'bad neighborhoods':

> There was a guest from Fujian Province making great noise in the neighbor's room at night ... Another guest from Fujian Province came to visit my neighbor and howled like a wild dog at midnight. He did not respond until I loudly rebuked him ... An unexpected visitor to my neighbor Wang talked loudly and ceaselessly. I tossed and turned in bed, frequently woken up. Suddenly this man began to abuse something occasionally with English language words inserted
>
> *(Xun 1959: 83).*

As Lu Xun went on to show, he detested his noisy neighbors:

> There is a man sick unto death; the next one gets their phonograph open; the opposite one is looking after children. Two persons laugh wildly upstairs with the sound of playing cards. A woman is crying for her dead mother on the boat of the river. We do not bear with the same grief. I just think they are too clamorous.
>
> *(Xun 1973: 100 – 101)*

Comments made by Schopenhauer suggest a similar problem with 'noise' in cities in Europe:

> At times, I am tormented and disturbed for a while by a moderate and constant noise before I am clearly conscious thereof, since I feel it merely as a constant increase in the difficulty of thinking, like a weight tied to my foot, until I become aware of what it is ... and it is precisely here that it is prevented by a noisy interruption. This is why eminent minds have always thoroughly disliked every kind of disturbance, interruption, and diversion, but above all the violent disturbance caused by din and noise.
>
> *(Schopenhauer 2000: 642 – 643)*

At the same time that Schopenhauer was writing *On Din and Noise*, other intellectuals like Thomas Carlyle (1795 – 1881), Charles Dickens (1812 – 1870) and John Leech (1817 – 1864) were also discussing modernity and 'noise' relating to the emergence of modern urbanism with reference to embodied, emotional and affective representations and experiences. In London, for example, Dickens depicted noises in the metropolis, such as clanging bells, cracking whips, clattering carriages, clamouring hawkers and cabmen and especially street music made by 'brazen performers on brazen instruments' (Picker 2000: 428). In a similar vein, Leech's dying words pointed to the problems with 'noise' that he had encountered, with him lamenting that 'rather than continue to be tormented in this way, I would prefer to go to the grave where there is no noise' (Frith 1891: 297).

From the point of view of these intellectuals in China and Europe, neighbourhood 'noise' was clearly closely related to class 'distinction' and 'a continuing struggle between refinement and vulgarity' in the emerging modern city (Bailey 1996: 60). For example, the population of London increased from 1,948,369 to 4,231,431 (a 117 per cent rise) between 1841 and 1891 (*Victorian London – Populations*, n.d.). However, as the *Report on the Sanitary Conditions of the Labouring Population of Great Britain* (1842) recorded 'the walls are only half-brick thick … materials are slight and unfit for the purpose' (Chadwick 1842: 347). As Thomas Carlyle complained early in 1824, the quality of buildings was contributing to 'noise' problems and social tension and individual conflicts:

> You are packed into paltry shells of brick-houses (calculated to endure for forty years, then fall); every door that slams to in the street is audible in your most secret chamber
>
> *(Carlyle 1824)*

Following on from such public concern in the late eighteenth and early nineteenth centuries representations and debate concerning the embodied, emotional and affective influence of 'noise' on city life have intensified since around 1900 and have been growing in parallel with the popularity of high-rise living. While high-rise living is often depicted as a symbol of economic development and social progress, academic research has often highlighted the challenges of 'noisy' living in high-density tall residential buildings: 'from the left and right comes mainly children noise, while children and furniture moving noise from the up and down neighbors … footsteps, doors and windows knocking' (Xie and Chen 1966: 170).

However, in an echo of even earlier historic depictions of building quality affecting urban soundscapes, commentators at the turn of the twentieth century in China argued that 'good-quality walls should be solid and at least 24cm thick. Thin walls are nonetheless often built with poor sound insulation by the real estate agency in order to enlarge the rooms' (Ma and Liu 2008: C17). Indeed, in a more recent survey of high-rise buildings in Chinese cities 95 per cent of floors did not reach the minimum of 75-decibel sound insulation required in the *Residential Building Code* published in 2006 (Ma and Liu 2008: C17). Moreover, some experts believe reinforced concrete that is used in high-rise buildings worsens the acoustic quality of private lives with 'Dutch engineers claiming that in reinforced concrete and iron framework buildings, the vibration of mechanical moving power could easily manifest itself' (Bijsterveld 2008: 164). Despite these infrastructural problems what is clear is that when upset and bothered by neighbourly noise many peoples's 'first response is to blame their upstairs neighbour while the truth is, mostly the real estate agency uses concrete floor-slabs between 10cm to 12cm with bad insulation quality and far from the national requirement' (Ma and Liu 2008: C17).

In this section I have highlighted embodied, emotional and affective problems relating to 'noisy' neighbours and presented examples of the ways in which academics, artists and writers in China, Europe and North American have consistently discussed this issue over the past few centuries and depicted 'noise' as an urban phenomenon (see Chapter 7, this volume). In these terms, 'noisy' neighbours can thus be understood as a defining feature of urban life since the development of the modern city in Europe in the late eighteenth century, and as we have seen such social and individual conflict appears to be an ever growing problem today in China with the increasing proliferation and popularity of high-rise living (see Chapter 4, this volume).

Noisy things

National guidelines suggest that noise inside civil residences should not rise above 50 decibels during the daytime and 45 decibels at night. Levels above 60 decibels are thought to be detrimental to health in numerous ways; for example, the chance of experiencing deafness is increased by 50 per cent if a person has long lived in an environment with noise levels consistently above 80 decibels (Liu 1994: 40). As well as hearing problems, 'noisy' urban life is being depicted as affecting eyesight, blood pressure and so on leading to dizziness, headache, insomnia, dreaminess, malaise, forgetfulness, fear, irritability, self-abasement, nervous disorder, and impaired female fertility (Geng 2001: 51).

While being a 'noisy neighbour' is still widely considered as socially unacceptable, TVs, audio systems, washing machines, vacuum cleaners, refrigerators, freezers, electric fans, air conditioners, ventilators, rice cookers, soybean milk makers, electric ovens, microwave ovens, are often sources of noise that lead to domestic conflict (see Chapter 9, this volume). Computers, laptops and mobile phones can also regarded as 'noisy' domestic appliances. Wang Min-an in his book *On Domestic Electric Objects* (2015) reflects on how 'noisy' contemporary life is inseparable from machines; and indeed 'noise' is thus central to daily life, family relationships and the social division of labour (see Chapter 6, this volume). In these terms, the 'home' has undergone a great change of 'industrialisation' which 'had brought complete audio systems into living rooms, effectively turning them into concert halls' (Bijsterveld 2008: 184; also see Chapter 10, this volume). TVs, for example, are often now the centre of family life in Chinese homes (see Wang 2015), while changes in plumbing technology ensure that 'while iron pipes were often adopted in the past, people now employ plastic pipes which are quite cheap but easily resonate and thus produce more noise' (Ma and Liu 2008: C17). Some customers have even complained that their washing machines sound like 'electric drills', or even 'a plane taking off', with online resources such as 'HEACN' (a special website for home electrical appliances) advising people how to weaken the noise from washing machines in order to reduce their own and neighbours' irritation (Mu 2015). Air conditioners and refrigerators have also been at the heart of conflicts between neighbours due to the impact of 'noise' on bodies, emotions and atmospheres (Wang 2015: 52).

Moreover, while not as audible 'noisy' computers and mobile phones have also had a significant impact on daily life, family relationship and social division of labour in contemporary cities. Ring tones, SMS messages, App notifications can all interrupt 'peaceful time and space, and disturb[s] the previous calmness' (Wang 2015: 122). Such 'noises' can be just as unexpected and sharp as a street whip-cracking that Schopenhauer denounced in the late eighteenth century as 'the very sting of the lash' (Schopenhauer 2000: 643). Also associated with contemporary digital technology is 'white noise'. However, in a letter to the Chinese translator of his book *White Noise*, Don DeLillo explains that:

> there are white noise devices that produce a kind of humming sound in which the intensity is the same at all frequencies. Such a device is designed to protect a person from other distracting or annoying sounds – street noises, aircraft, etc. This noise …, is 'uniform and white'.
>
> *(DeLillo 2001: 1).*

Don DeLillo's positive view on white noise is in stark contrast to David Harvey's (2014: 158) depiction of 'mass distraction' and 'privatopias' characterised by 'useless watching sitcoms, trawling the internet or playing computer games for hours on end'.

Noisy streets

At the end of the nineteenth century, traffic noise was described as 'the most cruel despot of our times' (Payer 2007: 784). Earlier Schopenhauer had berated the clip-clop of iron horseshoes on cobbled streets, harnesses rattling, car wheels clattering, carriages creaking and squealing, and as noted above, whips cracking, etc. Later, the electric streetcar, a symbol of modern metropolis, was also denounced for its 'whimpering howl' and 'infernal noise' (Payer 2007: 784). In combination with the introduction of motorised buses, car horns, bicycle bells, etc. it had been suggested that in the early modern 'since the creation of the world, no age has produced as much and such terrible noise' that has had embodied, emotional and affective influence on urban life (Payer 2007: 779).

Indeed, since the end of the nineteenth century, car ownership has been many people's dream or life goal. However, Xia Wang, who moved in 2007 to a high-rise building near to Xibeiwang in the Northern 5th Ring Road in Beijing, described night time in the new apartment as 'just like sleeping on the road'. Xia and her husband had taken to sleeping in the bathroom with her husband suggesting 'we quarreled a lot and we are scared of having a baby … Kids cannot bear such loud noise'(Ma and Liu 2008: C17). *The Report on China's Environmental Noise Pollution and Control* (2016) highlighted that traffic noise and increasing air traffic was a growing problem for urban residents.

Conclusion

In this chapter, I have advanced critical perspectives on soundscapes by briefly sketching out how neighbours, household appliances and traffic highlight the paradoxical nature of 'noisy' Chinese cities and the effect on bodies, emotions and atmospheres (see Gallagher and Prior 2014; Gandy and Nilson 2014; Revill 2016). While 'noisy' cities have been important signifiers of modernity and progress and 'noisy' people, things and streets can often make contemporary life in cities 'easier', more 'comfortable' and add to the 'buzz' of urban street life, in this chapter I have highlighted conflict and tension (see also Berman 1988). Understanding and responding to the ways in which 'noise' in cities is geographically and temporally significant will be a key challenge to advancing critical perspectives on urbanism both within and beyond China and highlights the importance of considering urban life with reference to bodies, emotions and atmospheres (see Chapters 4, 7 and 14, this volume).

16

CREATIVITY AND CHINESE URBANISM

The moral atmosphere of Lishui Barbizon

Jun Wang and Yan Li

Introduction

In a recent paper Gill and Pratt (2008) critique depictions of cultural workers as either a 'creative class of model entrepreneurs' or as part of 'a new 'precariat'. In a similar vein, cultural workers are often depicted as being inspired by positive emotions including love, self-expression and self-actualisation in order to undertake (im)material labour (Hardt 1999). Anxiety, fear and self-exploitation have also been discussed in the analysis of creative and cultural industries (Banks 2006; Oakley 2009). In Asia, empirical studies have described the precarious lives of cultural workers who not only endure economic hardship and insecurity but often with a backdrop of imposed political passivity (Chang 2009; Eberle and Holliday 2011) in the face of the uncontested construction of 'spectacular' cultural cities (Kong 2007; O'Connor and Xin 2006; Wang *et al.* 2016).

In order to offer new critical perspectives on creative-led initiatives in Chinese cities, in this chapter we focus on the affective labour of cultural workers in order to address two major questions that have emerged in recent academic writing. First, as Gill and Pratt (2008: 16 – 17) suggest 'how can [affect] ... be said to exist outside relations of power ... can it be claimed that affect is somehow outside the social? ... and how can affect be "normative" or disciplinary, binding us into structures and relations that may not be in our real interest?'. Second, is the important question regarding study of affect that enables the 'mobilizing potential of a place', that is, the ambient power of place where various human and non-human actors assemble in a way that particular moral value or social norms can be enacted, sensed, felt, and also reacted to (Allen 2006; Roberts 2012; Thrift 2007)? Such questions can be addressed through attention to relational interactions of body – environment including embodied and performative practices (Ahmed 2004b; Jayne and Leung 2014; Jayne *et al.* 2010).

Theorising affective atmosphere is of use here, in order to address diversity and heterogeneous body – environment interactions (Anderson 2009; Böhme 1993; McCormack 2008; Stephens 2015). We highlight moral atmosphere that is formed by multiple actors, not just top-down or bottom-up, but is diffusive and nebulous (Böhme 1993). With that argument in mind, we chart dynamic spatial-social relations – as ambient space or 'moody force fields' (Amin and Thrift 2013: 161) in the making of collective publics – to enrich our knowledge on 'what affect can and do' (Ahmed 2014: 38). Through the lens of affect, this chapter moves beyond 'common western assumptions' about Chinese governmentality (Ahmed 2014: 34).

Among the many experiments of cultural city making, the city of Lishui is an interesting case-study.[1] A small city situated next to a national natural reserve, Lishui is widely known for its association with the School of Lishui Barbizon – a Chinese counterpart of French Barbizon known for its *plein-air* painting. The identity of Lishui Barbizon, initially generated within the academic circle of artist-cum-academics, was adopted by city authorities in the 2000s. Materialising this imagined vision required significant labour including historical preservation of buildings and vistas; national search for artists who wanted to devote themselves to the moral values of Lishui Barbizon; and celebratory events and festivals to attract tourists all 'forming and formative processes of subjectivity' (Yang 2014a: 57).

With the city gaining a reputation for specific assemblages of emotions and feelings (Ahmed 2004b) it is not surprising that Lishui Barbizon has been infectious to many freelance artists. By deploying the concept of affective atmosphere, here we illustrate how moral values relating to Lishui Barbizon are not merely a top-down imposed state-sanctioned image, but are also felt, practised and embodied by subject-citizens who may respond with complex political passivity, and hegemonic and anti-hegemonic actions (see Chapter 14, this volume).

Affective atmospheres

An affective turn was first observed in humanities disciplines, and then in social sciences such as geography, sociology, social anthropology and so on (Anderson 2014; Clough 2008; Thrift 2004). The increasingly voluminous writing on affect can best be summarised through overlapping threads of thinking. In their edited volume the *Affective Theory Reader*, Gregg and Seigworth (2010: 5) identify 'two dominant vectors of affect study in the humanities': relating either to the work of theorists such as Silvan Tomkins, Spinoza or Deleuze. For Spinoza, the notion of affect straddles the divide between the power of mind and that of the body, which articulates the correspondence between 'the power to act', through embodied practices, and the power to be affected, through circulation of contagious emotions (Hardt 1999). Taking a different perspective to ask 'What is affect is?', Sara Ahmed (2004b) prioritises the question 'What can affect do?' In particular, her endeavour is to address through ethnography the functionality of affect, through which identities are constructed, negotiated and contested, producing collective and individual feeling and subjectivities (Ahmed 2004a).

Affect and moral image

Concern to understand how affect functions to delimit social, economic and political boundaries has also inspired numerous studies on the formative process of subjectivity making (Yang 2014b). Following Raymond Williams's (1977) conception of 'a structure of feeling', scholars have considered relations between affect and image, ideology, moral values, social norms and overall – politics. Ahmed is concerned with aggregated effects of the 'collective "us" and "others" through the contagious feature of affect. Emotion is felt and then reacted to, through this process, surfaces or boundaries are made: the "I" and the "we" are shaped by, and even take the shape of, contact with others' (Ahmed 2004b: 10). In this account, collective identity is effectuated by circulation of emotion. In other words, politics occur when individuals feel and align with people of likeness and form a community of the collective, through affective encounters such as alignment, identification, and appropriation (Yeoh 2015). As such, focus on 'how affect releases imagination, and intensifies connectivity between objects, people and events' (Yang 2014a: 45), allows us to explore the politics of image making, and how this is bound up in Chinese moralities.

In particular a focus on materialities to foreground the subject of atmosphere (Allen 2006; Anderson 2009; McCormack 2008; Stephens 2015; Thibaud 2003) has been particularly productive in elaborating body – environment interactions in 'mobilizing potential of a place' in making and shaping of collective publics (Amin and Thrift 2013: 161). In developing an 'aesthetic conceptualisation' of atmosphere, Böhme (1993) discusses the notion of 'aura' developed by Benjamin with respect to 'distance and respect' relating to original works of art. In these terms, atmosphere emerges from 'auras' that surround art-work to affect much broader settings, such as production of landscape, including elements such as water, sound, light, shade and foliage as playing a specific role in the production of atmosphere that 'seem to fill the space with a certain tone of feeling like a haze' (Böhme 1993: 114). Moreover, the notion of affective atmosphere suggests a dispersed and non-directional atmosphere which should not be limited to an object – subject moral system which is not necessarily binary but rather an assemblage of bodily and non-bodily experience (Stephens 2015). In a similar vein, ambient power can be perceived as a geographic constellation of buildings, people, technologies, and various forms of non-human life (Conradson and Latham 2007) including 'causal influences people have on themselves and each other, people sensed, perceived and therefore behave accordingly' (Wegner 2002). In this way affective atmospheres can be understood in two ways: first, by recognising subject/object distinctions; and second by acknowledging that atmosphere is subject to change, given that it is generated by bodily experiences of encounters with multiple (non)human actors (Anderson 2009).

In the remainder of this chapter we explore the diffuse, dispersed and non-directional features of atmosphere, not confined to a reading of ambient power but also with concern for how immaterial atmosphere affects and regulate people's behaviour, and how power emanates from (non)human actors. Drawing on this thinking, we examine how the affective atmosphere of Lishui Barbizon has 'taken hold and become infectious', felt as well as performed to forge identity and bounded community.

The image of Lishui Barbizon

Lishui is a city that hosts an artistic community called Lishui Barbizon. Founded on the celebration of natural landscapes that resembles the forest of Fontainebleau in France, the local institute of fine arts has endeavoured to establish its own identity by adopting ideas from the Barbizon School of French landscape. Rejecting the then prevailing classic style of painting of imagined religious scenery, Barbizon artists sought an accurate, impressionist presentation of actually existing scenes, turning their gaze to natural landscape and lives of farmers, gravediggers, woodsmen and poachers through *plein-air* painting, undertaken in the open air (Adams 1994). The Barbizon School of French landscape tended to be urban artists who moved to the countryside in the Forest of Fontainebleau in response to political turmoil in France, and through the growing political power and the taste cultures of the middle class and working class, and romantic political visions of rural life.

Interest in the French Barbizon School in China began in the early 1980s, when art-work was exhibited in Beijing and Shanghai. Impressed by the impressionist paintings of Fontainebleau, a group of artists at the Lishui University were quick to capture the potential of developing and distinguishing their own school based on their geographic place-luck – with their city of Lishui being located in a narrow valley near the Jiulong Mountain Natural Reserve. Running through the valley are three rivers called Ou, Min and Sai, which converge in the Liandu district. For those local artists the dense woods undercut with meadows, marshes, rivers and sandy clearings, resembled the landscapes of Fontainebleau and this inspired them to practise the French Barbizon principles which began formally with the first exhibition of the Lishui Barbizon school held in the China Academy of Fine Art in 1991.

For the Lishui Barbizon community, two major features have been distilled, and altered, from their French counterpart: First, artistic practice takes place through open air sketching and nostalgia for village life is preserved despite urbanisation. The second feature is a pursuit of immaterial life and endurance of hardship. For example, the university artists left their studios in town and set up their canvas near the river Ou, in their own language, 'to embrace the simple countryside lifestyles' (see Chapter 4, this volume). In 2005 and 2007, the government also organised several trips to Fujian and Guangdong to recruit trade-painters with a promise of transforming them into authentic artists.[2] In our study, we describe 'authorised artists' as academics-cum-artists in Lishui University whereas 'yet-to-be authorised artists' refers to these former trade-painters on a journey of imagined transformation. In 2009, there were forty-six trade-painting companies with two hundred and twenty 'yet-to-be authorised artists'. The Lishui Barbizon, exemplified by the two principles of *plein-air* painting and immaterial life, is underpinned by a moral image that requires artists' dedication to their embodied outdoor artistic practices and imagined artist lifestyle. In the following sections, we will illustrate how two moral principles are disseminated and embodied in forms of atmosphere and collective identity of Lishui Barbizon.

Being an authentic Barbizon artist

Ambient power: scenic park (jingqu) style of urbanisation

The very spot chosen that inspired Lishui Barbizon is Dagangtou in the Liandu district, where the two tributary brooks of Daxi and Songyin merge on their ways to the Ou River. About one mile away on the north bank is the heritage site of Tongji Weir where a ferry crosses between the two banks, therefore integrating the two places into one scenic spot, and which has been given the official title '*guyan huaxiang*' (ancient weir and hometown of artists, literally; artists' hometown, hereafter). In the middle of the tributary are two small islands with sandy clearings and clusters of high and dense bullrushes, making some vistas hidden whilst revealing others. Walking towards the riverside, you move from cement-paved streets to a pathway with pebbles of different sizes and then the view of Daxi brook comes into sight. On sunny days, two to three boats are often to be seen on the murmuring water, which rushes by until it is out of sights. Turn left and you get to The Old Street which whiles its way along the south bank of Daxi brook. The Old Street had been decaying. Like many traditional village buildings, these on The Old Street are black and white, with plain white sidewalls and dark grey roof tiles. The two facades are constructed by wooden frames and filled with removable wooden laths. Layers of protective oil reveal the natural colour of aged timber as dark brown. Other than the occasionally joyful cries of tourists the area is tranquil (see Chapters 4 and 15, this volume).

The organic and tranquil setting of the artists' hometown is the outcome of heritage conservation which sought to forge 'the aura of a utopia for intellectuals' (Interview 2015). Urbanisation in here follows an alternative path than many Chinese coastal cities. Such 'picturesque' urbanism, excluding the otherwise ceaseless urbanisation process that dominates elsewhere, freezing the city in a historical moment. underpins the local government's strategy towards being a 'cultural city' reliant on tourism.

In the first round of national territorial reconfiguration of administrative zones in the 1980s, Lishui was up-scaled to a county-level city after merging the Dagangtou town. It was then further up-scaled to a prefecture-level city, within the territory of a new urban district of Liandu which was established to govern the former Dagangtou town. Initially industrialisation was one major objective of that scalar reconfiguration, given that the 'simple equations of "city = industry" and "village = agriculture" have been deeply ingrained in the minds of all local officials' (Ma 2005: 483). In parallel, hundreds of small town-village-enterprises were established, often based on the 'Wenzhou model' which emerged in the same province

FIGURE 16.1 Lishui Old Street

Source: Jun Wang

where Lishui is located. However, the proximity to National Nature Reserve was one of the big constraints that limited Lishui's industrialisation. Instead, tourism became positioned as one key pillar industry from the late 1990s, and the so-called 'scenic park (*jingqu*)' style of urbanisation, which promotes a more integrated relation between nature and vernacular village life (Interview 2015; also see Chapter 4, this volume). The emergence of the national campaign to promote cultural cities in the early 2000s also gave the city justification to associate its tourism industry with culture-led regeneration. In 1999, Lishui defined itself as a 'hometown of photography' in conjunction with the National Association of Photographers. Since 2004, the International Lishui Festival of Photography has taken place every two years, inviting established photographers from all over the world to attend photo tours and to exhibit work in the city

Within this policy context the 'Alteration of the Overall Planning for Dagangtou Town 2005 – 2020' was formulated which specified the major principle of historic preservation. By that time, the old Dagangtou was decaying. Many of its former dwellers had moved to new homes to the south, leaving The Old Street deserted. The city government acquired the buildings, renovated them and allocated them to the Lishui Barbizon project. Building renovation was complemented by 'cleaning up' mountains, the brooks, the ancient weir, a two-storey Chinese style gazebo, old trees, to better present the landscape for painting.

The presence of two to three small boats with 'full sails' on murmuring water, at the same time every day 'is by design', answered one of our respondents, 'the city

FIGURE 16.2 'Best photo spot'

Source: Jun Wang

government paid the fishermen to have their boats there'. A more straightforward request to visitors is made by signboards strategically located here and there. For example, at the entrance of The Old Street, near a Ficus tree that is thousands of years old, visitors are told 'Best painting/photo spot'. Further along the brook, award-winning photos produced at that very spot were displayed, together with detailed instructions on recommended technical settings, such as the lens, aperture and shutter speed of visitors' camera. 'The natural landscape', the carefully preserved archaic buildings, the silence and occasional joyful cries, the soothing colours of light blue and dark brown forge the aesthetics of Lishui and correspondingly suggest behaviour that is required by such an aesthetics (Böhme 1993) and remind visitors to fulfil their duty of being an artist. 'This is the atmospheric power of place' (Allen 2006).

Lishui Barbizon as embodied and performative practice – us and them

The affective power of Lishui Barbizon was also supported in the 'casual conversations' artists tended to instigate with visitors regarding outdoor sketching and awards that had been won. Such performance of identity as belonging to the artistic community by conversations with 'others' highlights transpersonal intensity in the formative process of subjectivity, which is shaped by how we feel and how bodies respond to feelings (Ahmed 2004a). In other words, the bodies of artists practising *plein-air* painting itself as constitutive elements enhanced scenic spots. *Plein-air* painters appear in all advertisements, tourist guides' narratives, and blogs of tourists. Tourists seeking to claim cultural capital join freelance artists and students from fine art departments for sketching practice. The place therefore become a point of contact for a wide range of actors, from formal and yet-to-be-authorised artists, students, visiting artists and photographers, to tourists wishing to perform the artistic life. It is through embodied and performative practice that emotions circulate and travel between bodies to create the collective identity of us-and-them (see Chapter14, this volume).

Wang, a yet-to-be-authorised artist, is a passionate advocate of outdoor sketching. For him, *plein-air* painting is not only a means of sharpening his skill, but also an opportunity to

> learn from each other and get to know new friends … Just yesterday, I met two artists in the neighbouring village … I was near the river, but when they set up their easels in the villager's courtyard, I followed … you see, they know a better place to sketch.

Though he did not know their names, he could not hide his excitement at gaining the recognition or authorised artists, who purchased Wang's painting; 'this is a reward for my hard work … and certainly, it is a motive for work harder …'. Another yet-to-be-authorised artist from Fujian province proudly recalled the 'big names' he encountered while sketching together: 'the chairman of Dafen artists' association, chairman of Huizhou, I got to know all of them by sketching together … we all love particular spots, … and now we know each other very well'. These bodily performances of *plein-air* painting function to construct the 'surface' of the Lishui community, and at times enable let yet-to-be-authorised artists to challenge boundaries of belonging (Ahmed 2004a; Yeoh 2015). The tourist gaze is also a further important constitutive element of the Lishui Barbizon atmosphere. Upon arrival, tourists begin their journey – wandering around on The Old Street, holding their breath behind artists at work, visiting galleries, discovering 'authentic Barbizon art'. The noisy-ness, the frequent movements of their bodies, the flashlight of their cameras, all are 'out of tune' with the tranquil atmosphere (see Chapter 15, this volume).

During our fieldwork, we also met a cohort of fine art students undertaking summer school training. While the students were excited and playful at the beginning of their visit, switching from one spot to another they gradually settled themselves down and started to work in small teams. Aware of the admiring glances of tourists the students began to 'perform more professionally' as if to highlight Bohem's (1993) argument that the atmosphere is produced collectively by multiple bodily experiences, the heterogeneity of which is shaped by and in return reshapes the interaction between bodies and the surrounding settings. The aggregated effect of 'us' and 'them' produces the imagined community of Lishui Barbizon through performance, materialities and social relationship between bodies (Ahmed 2004a).

FIGURE 16.3 Us and them: encounters between painters and tourists

Source: Jun Wang

'Glory has its roots in immaterial life': life is hard; life of an artist is harder

Having interrogated affect, materialities and social relations, we now move to the second moral principle of Lishui Barbizon: pursuit of immaterial life, endurance of hardship that relate to feelings – positives ones like love and self-actualisation and negative ones such as shame, anxiety and fears that are embodied and performed in everyday life (Gill and Pratt 2008). In doing so we consider how atmosphere can be acted upon by moving bodies that are not simply singly directional and affected by a 'designed environment' influenced by politics and policy (Anderson 2009).

Celebrating plein-air painting

When we visited Lishui in 2015, the annual official art exhibition of Lishui Barbizon was taking place. Just days before, the General Secretary of the district government met Xi Jingping at a special group meeting of district governments. At that meeting, the Secretary was warmly greeted by Xi, who mentioned that he himself had visited Lishui Barbizon artists seven times during his term as General Secretary of Zhejiang Provincial Government. It is not surprising that Xi's words have been used by the General Secretary to promote Lishui Barbizon. When surrounded by international journalists, students, artists and flowers, flags and banners the Secretary praised the artists for their life-long passion and devotion to portraying the natural landscape of their hometown, and expressed his gratitude to the artists and the city of Lishui Barbizon that impressed the leader of China.

This event is just one example of the ways in which the district government have been celebrating *plein-air* painting, such as during annual sketching festivals. Held for the previous six years the sketching festivals were open-field events from April to September, when academics-cum-established artists, students, Lishui Barbizon artists, yet-to-be authorised artists and amateurs gather – a fiesta for art for the city. Indulging in a joyful atmosphere, Mr. Zhan Weike, one of the founders of the School and now a 'star artist' of Lishui Barbzon, recalled the 'old and harsh days' when they first adopted *plein-air* painting to follow the French pioneers. He pointed to the 'days after days and years after years' that the artists stayed in the mountains to practise outdoor painting. As Zhan reiterated, 'only natural landscape, only out-door painting, we must stick to these two principles'. He highlighted how outdoor sketching kept the artists' feet deeply rooted in the 'natural landscape' nurturing competence in discovering, sensing, and expressing beauty on canvas through endurance of hardship. 'In my cohort', recalled Zhan, 'the everlasting pursuit of art pushed us to try all kinds of means to improve ourselves … It was a long and lonely journey while artists in other places shifted from one genre to another genre to pursue material wealth … The glory has its root in immaterial life!'

Practising immaterial life

Indeed, such hardship was vivid in the memory of those formal university artists, who are now are enjoying rewards in both market value and fame. Hardship nonetheless remained an everyday reality for yet-to-be-authorised artists in Lishui. One glance at the homes of those aspirational artists is revealing. In most of their bedrooms there is only very basic furniture such as a self-made bed, a chair, stove and computer. In contrast, the ground-floor galleries were carefully designed and arranged. Almost every inch of the three walls was covered by paintings of different sizes, clustered by genres or theme. In almost all of the galleries, siting at the centre of the room was one large Chinese wood-carved tea table, with a complete Chinese tea set and an electric kettle on its top, surrounded by wood-carved chairs. During visits to galleries, patrons are invited to drink tea and to sit down and partake in conversation. While there is generally no implications of 'selling', the aesthetic and monetary value of the paintings was often mentioned.

It was during such an occasion that we met Wang, a self-claimed representative of the street. At the time of our visit, he sojourned at a studio owned by an academic-cum-artist through 'some guanxi'. In the day-time, he looked after that gallery or any other gallery when required by their owners who had gone out sketching. If time allowed, he also spent time painting outside himself. While we were having tea, Wang introduced work by the gallery owner, whose success must be ascribed to the long history of practising Barbizon principles. Wang then directed us to his own paintings. Wang moved to Lishui around ten years ago, to pursue his dream of being a successful professional artist. After an 'artistic pilgrimage' that took Wang

FIGURE 16.4 'Yet-to-be authorised' artist Wang in the studio

Source: Jun Wang

to Beijing, Fuzhou, Shenzhen, Lijiang, and then all the way to the Tibetan Plateau to 'gain experience and understand the market of painting', he found his way to Lishui. Recalling his experience:

> I joined the army when I was 20 and since then I've been working as artist, janitor, security guard, village secretary, construction worker ... do you know I have tried tens of jobs all these years? But if I'm not starving, I would go back to learn painting.

Nurtured by his experience of hardship, Wang believed that he would become a better artist through day-by-day struggle in order to follow in the footsteps of the many famous painters who had only gained recognition during old age – endurance was key to an 'immaterial life'. Indeed, many of the artists that we interviewed often expressed mild annoyance at our questions about their economic situation: 'It is immaterial ... there is no such a thing as a material artist!' As another yet-to-be authorised artist explained to us in his studio,

> This is the place for artists ... so tranquil, so zen, it purifies your mind ... you see, the brook is just outside my door, it always calms me down and leads me into the right mood of painting ... so artists here are not like those outside, we don't care so much about money ... life is hard, life of an artist is harder.

'The glory has its roots in immaterial life', was a slogan that seemed to integrate positive emotions of being an 'authentic Barbizon artist' in order to manage anxiety and personal risk as inevitable on the path towards success. Such evidence highlights Ahmed's (2004a) and Yang's (2014a) contentions regarding relational 'objectivities of emotion' that are felt, understood, interpreted and performed as part of everyday life in Lishui in a manner that creates affective atmospheres emanating from bodies, but which is not reducible to them. Taken up and reworked in lived experience – becoming part of feelings and emotions that may themselves become elements within other atmospheres (Anderson 2014) such structure of feeling is extended towards a circulation of affect which is under constant reworking during indeterminate encounters with a constellation of bodies and non-bodies. It is exactly because performative practices of subject-bodies are constitutive elements of atmosphere, circulation of affect, and through performative power that both both authorised and yet-to-be authorised artists commit to ceaseless behaviour of 'self-exploitation' and 'passive' acceptance of the creative city script.

Conclusion

In this chapter we have examined the construction of Lishui Barbizon as a cultural city constituted through affective atmosphere and moral governance. In doing so we have challenged binary readings of state – society relations and the structure-agency in Chinese urbanism. First, we have shown how the Chinese state does not govern solely through cohesive force, but through constructing moral-value-laden environments and forging affective power. However, in moving beyond a one-directional interpretation of state-sanctioned affective imagery, we have urged the importance of a diffusive reading of power relations. For example, we have highlighted how affective atmospheres can be enabled, by (re)arranging 'natural' and human landscapes, colour, light, sounds, and mutually affective bodies (Böhme 1993). Such focus points not to the importance of single sources that produced a collective feeling generated to micro subjects through central engineering, but as flowing through a heterogeneous assemblage of (non)human actors and reproduced through a circulation of affect. Consideration of collective diffusive, borderless, interactive patterns, instead of a clear-cut depictions of 'the state' and 'civil society', 'resistance' and 'passivity' highlights the importance of work that seeks to advance complex and sophisticated theoretical and empirical critical perspectives on Chinese urbanism (see for exmple, Stephens 2015; Wang and Chen, forthcoming; and Chapter 14, this volume).

Notes

1 Data used in this study was collected in semi-structured interviews and on-site observations during fieldwork in March 2015 and November 2016. We also draw on the personal accounts of artists and public representation of Lishui by painters on social media, and also through conversations on WeChat (Chinese social media) with interlocutors living and working in Lishui.

2 Trade-painters mainly work on copying masterpieces on an assembly line. Please see Wang, J. (2016) 'Worlding through Shanzhai: the evolving art cluster of Dafen in Shenzhen, China', in *Making Cultural Cities in Asia: Mobility, Assemblage, and the Politics of Aspirational Urbanism*, edited by J. Wang, T. Oakes and Y. Yang. London: Routledge, pp. 129 – 44.

Acknowledgements

This research was supported by grants awarded by the Research Grants Council of the Hong Kong Special Administrative Region, China (Project No. CityU 247713 and CityU 21613815).

17

AFTERWORD

Critical Chinese urbanism for the twenty-first century

Mark Jayne

Wow! Did you enjoy yourself? What an inspiring collections of essays! The theoretical, empirical and methodological terrain covered in this book clearly highlights the strengths of contemporary work being undertaken by Chinese urban scholars and scholars of urban China. The arguments made, and debates explored by each contributor showcase the opportunities offered by developing critical perspectives on Chinese cities, not only as an important project in itself but also because of the fruitful avenues that are opened up to advance international urban theory more broadly. In these terms, this book is an important milestone. The chapters are a snapshot of the advances made, the successes achieved. However, as the authors highlight there is still much hard work, thinking, writing and research that needs to be done in order to ensure that Chinese urbanism not only engages with and challenges, but takes a lead in advancing critical- and cutting-edge theoretical debate in the twenty-first century.

As such, while it is important to celebrate the critical perspectives advanced by each contributor, as well as the work of urbanists and other theorists who focus on China that are referenced throughout the pages of this book, it is also important to take a moment to reflect on the challenges at hand. While championing theoretically driven writing and research we must acknowledge for example, that contemporary Chinese urban scholarship remains overwhelmingly characterised by 'empirical studies and econometrical modeling ... [and hence] scholarship in urban China studies [has been] ... dominated by a positive approach and generally lacked nuanced analysis and theoretical debates' (He and Qian 2016: 463). At present, it is thus fair to say that writing which offers critical and theoretically driven perspectives remains marginal in comparison to more empirically and policy-led research in China – although these are of course not mutually exclusive approaches, they can interpenetrate and complement each other. However, in seeking to address and overcome this imbalance we should nonetheless be mindful of Hubbard's (2006) warning against the simple 'out with the old in with the new approach' where established lexicons and orthodoxies are replaced with neologisms. What is important then is that a more equitable blend of theoretically driven critical perspectives exist in parallel, or even, preferably, in productive tension or dialogue with the strong traditions of empirically focused and policy-led research in order to enable theoretical, empirical and methodological innovation. Of course, if critical Chinese urban thinking and theorists are to take centre stage in leading contemporary and future international debates then political, intellectual, institutional and financial support and recognition within and beyond universities in China and elsewhere in the world is a necessity.

In foregrounding the role of urban theory within Chinese scholarship, as mentioned in the introductory chapter, it is vitally important not to ignore the significant critical writing by Chinese scholars, not published in English language journals and books, or not directly engaging with urbanism *per se*. This writing and research offers rich and detailed resources to advance our understanding of political, economic, social and cultural life in China which includes a diverse range of topics, case studies and contexts. This assertion raises challenges of its own of course; first, to ensure that when authors are writing in/about cities that urbanism and critical perspectives are at the heart of the writing and research. 'The city' must become centre stage in Chinese scholarship in disciplines across the humanities, social science, health and medical sciences. Second, there is clearly also a need for international scholars to take seriously writing in Chinese and the work of those who develop different approaches to 'doing theory' that do not implicitly draw on European and North American traditions of knowledge production, genealogies of ideas, key thinkers, writing conventions and so on. It is thus an important and fascinating project to critically reflect on the place of Chinese urban theory in relation to the geographically uneven foundations of contemporary urban scholarship. Such work is vital to advance an ontological and epistemological agenda that enables us to explore how historical and contemporary Chinese scholarship contributes to understanding urban spatiality with distinctively situated social relations, practices and institutions whilst recognising global processes and a host of urban imaginaries from around the world (see Edensor and Jayne 2011).

Such theoretical impulses are vital for advancing critical perspectives on Chinese urbanism. For example, on a recent flight back from a research trip to Shenzhen I settled down to watch *Citizen Jane: Battle for the City*, which detailed the life, academic work and citizen activism of Jane Jacobs, including an exploration of perhaps her most famous writing, *The Death and Life of Great American Cities* (1961). While this was an enjoyable watch, I was dismayed as an urbanist and more specifically an as urban geographer, by the 'throwaway' comments made towards the end of the film that suggested mistakes and failings made in cities in North America and Europe from the 1950s onwards are now being replicated and exaggerated in China on an unprecedented scale. My concern at such uncritical commentary is of course, not a suggestion that there are not very real problems, or significant concerns about the rapid expansion of Chinese cities over the past few decades, and worries about the scale and speed of future planned growth. However, critical Chinese urbanism for the twenty-first century must be confident in outlining theoretical, empirical and methodological resources which critique and challenge such problematic and overly simplistic application of 'already existing' (from elsewhere) frames of thinking being applied to cities in China whether those come from academic, political, policy or popular contexts.

It will be clear from reading the chapters in this book that contemporary critical writing on Chinese urbanism is constituted by a diverse range of theoretical perspectives, empirical case-studies and methodological approaches which when read together are in a strong position to advance that agenda. Urban theory is the unique contribution that academics make to understanding cities both within and beyond universities (see Jayne and Ward 2017). However, urban theory is not pre-configured, its boundaries are made and remade, reworked, subject to contestation and refinement from inside and from outside universities. Chinese urbanists must continue to respond to the challenge of asserting how critical perspectives capture complexity and diversity, draw on rich intellectual traditions of creative and innovative thinking from around the world, and keep pace with the ongoing development of the urban experience (see Ward 2009; Connell 2007 Robinson 2006). This challenge should offer inspiration to those of you who are determined to advance understanding of cities in China and to locate Chinese urban theory at the heart of international research agendas.

As I sit writing these concluding remarks, in the small city of Cardiff, Wales having thoroughly enjoyed reading over the chapters of this book one more time, I find myself reliving the excitement, confusion, bemusement, amusement and thrilling sights, sounds, tastes, smells and atmospheres of Chinese cities. I find myself missing my Chinese friends and colleagues despite WeChat keeping us connected, I am hungry for Chinese food, thirsty for the fabulous locally brewed craft beers as well as my favorite Mao-tai wine, I want to experience again the sensations and atmospheres of street-life – Chinese cities are indeed awe-inspiring. However, while missing China, I am also aware that Chinese urbanism tends to surrounds me wherever I am. In Wales, local Chinese residents are now the third largest ethnic group in the country. Across the globe, there can be little doubt about the important ways that such long-standing Chinese communities, along with the presence of more recently increasingly visible groups such as students and tourists add vitality to cities. Mo Bikes can now be found on the streets of Manchester (UK), 'Chinatowns' continue to be cosmopolitan and vibrant urban spaces in many cities around the world; stories about China seem to appear in the news daily; references to Chinese everyday life are proliferating in popular cultural forms and practices.

The chapters of this book celebrate both past and contemporary ground-breaking theoretical work and new and exciting research, debate and argument relating to Chinese urbanism. Critical Chinese urbanism for the twenty-first century needs to interrogate the similarities, differences, connectivities, territorialities and relationalities of urban life. Chinese theorists and scholars of urban China must thus inspire and challenge within and beyond universities by theorising, researching and writing about cities in China and by paying attention to the ways in which Chinese urbanism exists in cities throughout the world.

BIBLIOGRAPHY

Adams, S. (1994) *The Barbizon School and the Origins of Impressionism*, Paris: Phaidon Press.

Adey, P. (2010) *Mobility*, New York/London: Routledge.

Adom, K. (2014) 'Beyond the marginalization thesis: an examination of the motivations of informal entrepreneurs in Sub-Saharan Africa: insights from Ghana', *Entrepreneurship and Innovation*, 15: 113–25.

Agyeman, J., Bullard, R. and Evans, B. (2002) 'Exploring the nexus: bringing together sustainability, environmental justice and equity', *Space and Polity*, 6: 77–90.

Ahmed, S. (2004a) 'Collective feelings: or, the impressions left by others', *Theory, Culture and Society*, 21(2): 25 . 42.

Ahmed, S. (2004b) *The Cultural Politics of Emotion*, Edinburgh: Edinburgh University Press.

Ahmed, S. (2014) 'Starting and startling', in Yang, J. (ed.), *The Political Economy of Affect and Emotion in East Asia*. Oxford and New York: Routledge, 24–39.

Alexander, C., Gregson, N. and Gille, Z. (2013) 'Food waste', in Murcott, W., Belasco, G. and Jackson, P. (eds) *The Handbook of Food Research*, London: Blomsbury, 471–484.

Allen, J. (2006) 'Ambient power: Berlin's Potsdamer Platz and the seductive logic of public spaces', *Urban Studies*, 43(2): 441–455.

Allen, J. (2000) 'On Georg Simmel: proximity, distance and movement', in Crang, M. and Thrift, N. (eds) *Thinking Space*, London and New York: Routledge, 54–70.

Amin, A. (2008) 'Collective culture and urban public space', *City*, 12(1): 5–24.

Amin, A. and Graham, S. (1997) 'The ordinary city', *Transactions of the Institute of British Geographers*, 22(4): 411–429.

Amin, A. and Thrift, N. (2013) *Arts of the Political: New Openings for the Left*, New York: Duke University Press.

Amin, A. and Thrift, N. (2002) *Cities: Re-Imagining the Urban*, Cambridge: Polity.

An, H. (1991) *The History of Chinese Gardens*, Shanghai: Tongj University Press.

Anagnost, A. (2004) 'The Corporeal Politics of Quality *(Suzhi)*', *Public Culture*, 16: 189–208.

Anderson, B. (ed.) (2014) *Encountering Affect: Capacities, Apparatuses, Conditions*, Aldershot: Ashgate.

Anderson, B. (2009) 'Affective atmospheres', *Emotion, Space and Society*, 2(2): 77–81.

Anderson, B. and Harrison, P. (2010) *Taking-Place: Non-Representational Theories and Geography*, Aldershot: Ashgate.

Anderson, B. and Tolia-Kelly, D. (2004) 'Matter(s) in social and cultural geography', *Geoforum*, 35: 669–674.

Andres, L. (2013) 'Differential spaces, power hierarchy and collaborative planning: A critique of the role of temporary uses in shaping and making places', *Urban Studies*, 50(4): 759–775.

Andres, L. and Golubchikov, O. (2016) 'The limits to artist-led regeneration: creative brownfields in the cities of high culture', *International Journal of Urban and Regional Research*, 23(4): 757–775.

Appadurai, A. (1996) *Modernity at Large: Cultural Dimensions of Globalization*, Minneapolis: University of Minnesota Press.

Arnould, E. J. and Price, L. L. (1993) 'River magic: extraordinary experience and the extended service encounter', *Journal of Consumer Research*, 20(1): 24–45.

Attali, J. (1985) *Noise: The Political Economy of Music*. Trans. Brian Massumi. Minnesota: University of Minnesota Press.

Atkinson, R. and Easthope, H. (2009) 'The consequences of the creative class: the pursuit of creativity strategies in Australia's cities', *International Journal of Urban and Regional Research*, 33(1): 64–79.

Atsushi, K. (1979) *The History of Hainan Island* (Translated from Chinese by Zhang, X). Taipei: Xue Hai Press.

Bachelard, G. (1964) *The Poetics of Space*, New York: Orion Press.

Bahl, S. and Milne, G. R. (2010) 'Talking to ourselves: a dialogical exploration of consumption experiences', *Journal of Consumer Research*, 37(1): 176–195.

Bailey, P. (1996) 'Breaking the Sound Barrier: A Historian Listens to Noise', *Body and Society*, 2(2): 49–66.

Balibrea, M. P. (2001) 'Urbanism, culture and the post-industrial city: challenging the "Barcelona model"', *Journal of Spanish Cultural Studies*, 2(2): 187–210.

Banks, M. (2006) 'Moral economy and cultural work', *Sociology*, 40(3): 455–472.

Baron, L. (1982) 'Noise and degeneration: Theodor Lessing's crusade for quiet', *Journal of Contemporary History*, 17(1): 165–178.

Barth, J., Lea, M. and Li, T. (2012, October) *China's Housing Market: Is A Bubble About To Burst?* Milken Institute,

Baudrillard, J. (1983) *Symbolic Exchange and Death*, London: Verso.

Bauman, Z. (1997) *Postmodernity and its Discontents*, New York: New York University Press.

BAUPD (2013) *The Evaluation Report on the Implementation of Urban Planning and Green Belt Policy in Beijing's City Proper* [Beijingshi zhongxin cheng lv'ge diqu guihua shishi pinggu], Beijing: Beijing Academy of Urban Planning and Design.

Beaverstock, J. G., Lorimer, H., Smith, R. G., Taylor, P. J. and Walker D. R. F. (1999) 'A roster of world cities', *Cities*, 16: 445–458.

Beer, A., Delshammar, T. and Schildwacht, P. (2003) 'A changing understanding of the role of greenspace in high-density housing: A European Perspective', *Journal of Built Environment*, 29(2): 132–143.

Beijing Archives (1958) *No. 1–5-253: Report of Beijing Municipal Party Committee to the Central Committee of CPC on the preliminary urban plan of Beijing* [Shiwei guanyu beijing chengshi guihua chubu fangan xiang zhongyangde baogao], edited by Beijing Municipal Party Committee. Beijing: Beijing Archives.

Beijing Municipal Government (1999) 'Protection and control planning of historic and cultural preservation areas in Beijing old city' *(Beijing jiucheng lishi wenhua baohuqu baohu he kongzhi fanwei guihua)*', Available at: http://www.bjww.gov.cn/2005/8-3/14310.html (Accessed 16 March 2017).

Beijing Municipal Government (2002) 'Protection planning of 25 historic and cultural preservation areas in Beijing old city' *(Beijing jiucheng 25 pian lishi wenhua baohuqu baohu guihua)*', Available at: http://www.bjww.gov.cn/2005/8-3/14366.html (Accessed 16 March 2017).

Beijing Youth Daily (2014) 'Dongguan under antiprostitution crackdown', Available at: http://epaper.ynet.com/html/2014-02/19/content_41612.htm?div=-1 (Accessed 2 June 2015).

Beijing Youth Daily (2000) 'Beijing deloys the 2008 Olympic bid [Beijing bushu 2008 nian shen'ao]',Available at: http://www.people.com.cn/GB/channel1/10/20000818/192554.html Last modified 18 August 2000 (Accessed 15 April 2017).

Belk, R. (1988) 'Possessions and the extended self', *Journal of Consumer Research* 15(2): 139–168.

Bell, D. and Jayne, M. (2009) 'Small cities? Towards a research agenda', *International Journal of Urban and Regional Research*, 33(3): 683–699.

Bell, D. and Jayne, M. (eds) (2006) *Small Cities: Urban Experience Beyond the Metropolis,* London: Routledge.

Bell, D. and Jayne, M. (2001) '"Design-led" urban regeneration: A critical perspective', *Local Economy*, 18(2): 121–134.

Benjamin, W. (2006) *Berlin Childhood Around 1900*, Cambridge, MA: Belknap Press.

Benjamin, W. (1999) *The Arcades Project* (translated by H. Eiland), Cambridge, MA: Harvard University Press.

Bennett, K. (2006) 'Kitchen drama: Performances, patriarchy and power dynamics in a Dorset farmhouse kitchen', *Gender, Place and Culture,* 13(2): 49–56.

Benton-Short, L. and Short, J-R. (2008) *Cities and Nature*, London: Routledge.

Berman, M. (1988) *All That is Solid Melts into Air: the Experience of Modernity*, Harmondsworth: Penguin Books.

Bhabha, H. (1994) *The Location of Culture*, London: Routledge.

Bhatti, M. and Church, A. (2004) 'Home, the culture of nature and meanings of gardens in late modernity', *Housing Studies*, 19(1): 37–51.

Bhatti, M. and Church, A. (2001) 'Cultivation natures: homes and gardens in Late Modernity', *Sociology*, 35(2): 365–383.

Bijsterveld, K. (2008) *Mechanical Sound: Technology, Culture, and Public Problems of Noise in the Twentieth Century*, Cambridge, MA: MIT Press.

Biles, J. J. (2008) 'Informal work and livelihoods in Mexico: getting by or getting ahead?' *Professional Geographer*, 60: 541–555.

Bissell, D. (2007) 'Animating suspension: waiting for mobilities', *Mobilities,* 2(2): 277–298.

Bissell, D. and Fuller, G. (2011) 'Stillness unbound', in Bissell, D. and Fuller, G. (eds), *Stillness in a Mobile World*, New York, NY: Routledge, 1–17.

BLGCGB (2002) *BLGCGB-2002-No.11: Notice on Making an Uniform Arrangement of Remaining Construction Land Plots in the Green Belt Area* [Jingzongzhifa 2002–11 hao: Guanyu tongyi anpai benshi lvhua geli diqu fanwei nei shengyu jianshe yongdi de tongzhi]m Beijing: Beijing Leading Group of Constructing the Green Belt Area.

BLGCGB (2000a) *BLGCGB-2000-No.2: Minutes of the Meeting on the Implementation of 'All-in-one' Approval for Green Industrial Projects in Beijing's Green Belt Area* [Shizongzhihui 2000–2 hao: Guanyu benshi lvhua geli diqu jingyingxing lvse chanye xiangmu shixing "yitiaolong" shenpi youguan wenti de huiyi jiyao], Beijing: Beijing Leading Group of Constructing the Green Belt Area.

BLGCGB (2000b) *BLGCGB-2000-No.7: Minutes of the Meeting on Further Accelerating the Approval Progress of Construction Projects in the Green Belt Area* [Jingzongzhihui 2000–7 hao:

Guanyu jinyibu jiakuai benshi lvhua geli diqu jianshe xiangmu shenpi jindu youguan wenti de huiyi jiyao], Beijing: Beijing Leading Group of Constructing the Green Belt Area.

Block, A. and Farías, I. (eds) (2016) *Urban Cosmopolitics: Agencements, Assemblies, Atmospheres*, London: Routledge, 1–22.

BMCUP (1992) 'Beijing urban master plan (1991–2010)', *Beijing Municipal Commission of Urban Planning*, Last modified December 1992. Available at: http://bj.leju.com/o/2002-05-22/11536.html (Accessed 7 February 2017).

BMCUP (ed.) (1987) *Collected information on Beijing Urban Construction works since 1949 (Volume 1: urban planning)* [Jianguo yilaide beijing chengshi jianshe ziliao: diyi juan - Chengshi guihua], Beijing: Editorial Board for Historical Records of Beijing Construction Works, BMCUP.

BMG (2002a) *BMG-2002-No.4: Notice on Strengthening the Administration of State-owned Land Assets and Establishing Land Reservation System* [Jingzhengfa 2002–4 hao: Guanyu jiaqiang guoyou tudi zichan guanli, Jianli tudi chubei zhidu yijian de tongzhi], Beijing: Beijing Municipal Government.

BMG (2002b) *BMGGO-2002-No.33: Relevant Provisions on the Suspension of Close-door Negotiations in Leasing State-owned Land Use Rights* [Jingzhengbanfa 2002 di 33 hao: Guanyu tingzhi jingyingxing xiangmu guoyou tudi shiyongquan xieyi churang de youguan guiding], Beijing: General Office of Beijing Municipal Government.

BMG (2001a) *Ordinance on Managing Land Replacement in Beijing's Green Belt Area* [Beijingshi lvhua geli diqu tudi zhihuan guanli banfa], Beijing: Beijing Municipal Government.

BMG (2001b) *BMGGO-2001-No.31: Opinions on Speeding up the Transformation of Old Villages and the Construction of New Villages in the Green Belt Area* [Jingzhengbanfa 2001–31 hao: Guanyu jiakuai benshi lvhua geli diqu jiucun gaizao he xincun jianshe de shishi yijian], Beijing: General Office of Beijing Municipal Government.

BMG (2000a) *BMG-Meeting-2000-No.34: Minutes of the Meeting of BMG on Speeding up the Construction of the Green Belt Area* [Jingzhenghui 2000–34 hao: Beijing shizhengfu guanyu jiakuai lvhua geli diqu jianshe youguan wentide huiyi jiyao], Beijing: Beijing Municipal Government.

BMG (2000b) *BMG-2000-No.12: Ordinance on Speeding up the Construction of Green Belt* [Jingzhengfa 2000–12 hao: Guanyu jiakuai benshi lvhua geli diqu jianshede yijian], Beijing: Beijing Municipal Government.

BMG (1999) *Minutes of Mayor's Office Meeting on 29 September 1999* [1999 nian jiuyue ershijiuri shushing bangonghui huiyi jiyao], Beijing: Beijing Municipal Government.

BOCOG (2001) *Bidding Statement of the Delegate of The Beijing Organizing Committee for the Games of the XXIX Olympiad* [Beijing shen'ao daibiaotuan chenshu baogao quanwen], China Internet Information Centre, Available at: http://www.china.com.cn/chinese/kuaixun/44673.htm Last modified 13 July 2001. (Accessed 25 March 2017)

Böhme, G. (1993) 'Atmosphere as the fundamental concept of a new aesthetics', *Thesis Eleven*, 36(1): 113–126.

Boren, T. and Young, C. (2012) 'Getting creative with the "creative city"? Towards new perspectives on creativity in urban policy', *International Journal of Urban and Regional Research*, 37(5): 1799–1815.

Bourdieu, P. (1987) *Distinction: A Social Critique of the Judgment of Taste* (trans. Richard Nice.), Boston, MA: Harvard University Press.

Braester, Y. (2010) *Painting the City Red: Chinese Cinema and the Urban Contract*, Durham, NC: Duke University Press.

Bray, D. (2006) 'Garden Estates and Social Harmony: A Study into the Relationship between Residential Planning and Urban Governance in Contemporary China', Paper presented at the Third Annual Conference of China Planning Network, Beijing, China, 14–16 June.

Bray, D. (2005) *Social Space and Governance in Urban China: The Danwei System from Origins to Reform*, Stanford, CA: Stanford University Press.

Brechin, S. and Kempton, W. (1994) 'Global environmentalism: a challenge to the postmaterialism thesis?', *Social Science Quarterley*, 75(2): 245–269.

Brenner, N. (ed.) (2014) *Implosions/Explosions: Towards a Study of Planetary Urbanization*, Berlin: Jovis Verlag.

Brenner, N. (2009) 'What is critical urban theory?', *City*, 13: 198–207.

Brenner, N. (2004) 'Urban governance and the production of new state spaces in Western Europe, 1960–2000', *Review of International Political Economy*, 11: 447–488.

Brenner, N. (1999) 'Globalisation as reterritorialisation: the re-scaling of urban governance in the European Union', *Urban Studies*, 36(3): 431–451.

Brenner, N. and Schmid, C. (2014) 'The "urban age" in question', *International Journal of Urban and Regional Research*, 38: 731–755.

Brenner, N., Madden, D. J. and Wachsmuth, D. (2011) 'Assemblage urbanism and the challenges of critical urban theory', *City*, 15(2): 225–240.

Brenner, N., Marcuse, P. and Mayer, M. (eds) (2011) *Cities for People, Not for Profit: Critical Urban Theory and the Right to the City*, London: Routledge.

Brents, B. G., Jackson, C. A. and Hausbeck, K. (2010) *The State of Sex: Tourism, Sex and Sin in the New American Heartland*, New York and London: Routledge.

Brighenti, A. M. (2016) *Urban Interstices: The Aesthetics and the Politics of the In-between*, London: Routledge.

Brødsgaard, K. E. (2009) *Hainan: State, Society and Business in a Chinese Province*, Abingdon: Routledge.

Bromley, N. (2004) 'Flowers in her bathtub: boundary crossings at the public-private divide', *Geoforum*, 36: 281–296.

Bromley, R. (1990) 'A new path to development? The significance and impact of Hernando De Soto's ideas on underdevelopment, production, and reproduction', *Economic Geography*, 66: 328–348.

Broudehoux, A. M. (2004) *The Making and Selling of Post-Mao Beijing*, New York: Routledge.

Buckley, S. (1996) 'A guided tour of the kitchen: seven Japanese domestic tales', *Environment and Planning D: Society and Space*, 14: 441–461.

Burgess, E. (1925) 'The growth of the city: an introduction to a research project', in Park, R., Burgess, E. and McKenzie, E. (eds) *The City*, Chicago: The University of Chicago Press.

Burgess, E. (1923) 'The growth of the city: an introduction to a research project', *Publications of the American Sociological Society*, 18: 86–97.

Burrell, K. (2008) 'Materialising the border: Spaces of mobility and material culture in migration from post-socialist Poland', *Mobilities*, 3(3): 353–373.

Butcher, M. (2010) 'From "fish out of water" to "fitting in": the challenge of re-placing home in a mobile world', *Population, Space and Place*, 16(1): 23–36.

Cai, X., Gan, Q. and Zhang, C. (2004) ,The spatial distribution and impact factor of the foodscape in Guangzhou' (in Chinese). *Social Scientist* (2): 95–98. [蔡晓梅, 甘巧林, 张朝枝. (2004). 广州饮食文化景观的动检特征及其形成机理分析. 社会科学家 (2): 95–98.]

Cai, X., Xiong, W. and Situ, S. (2006) 'Eating in Guangzhou: The cultural connotation and underlying causes', (in Chinese). *Tropical Geography* 26(2): 192–196. [蔡晓梅, 熊伟, 司徒尚纪. (2006). "食在广州"的文化内涵与成因分析. 热带地理 26(2): 192–196.]

Cameron, S. and Coaffee, J. (2005) 'Art, gentrification and regeneration – From artist as pioneer to public arts', *European Journal of Housing Policy*, 5(1): 39–58.

Cangiani, M. (2011) 'Karl Polanyi's institutional theory: market society and its "disembedded" economy', *Journal of Economic Issues*, 45(1): 177–198.

Cant, S. G. (2012) 'In close encounter: the space between two dancers', in Paterson, M. and Dodge, M. (eds) *Touching Space, Placing Touch,* Aldershot: Ashgate, 211–230.

Cao, J. and Chen, Z. (1997) *Leaving the Ideal Castle: Research on China's Danwei Phenomenon,* Shenzhen: Haitian Press

Caprotti, F. (2014) 'Eco-urbanism and the eco-city, or, denying the right to the city?' *Antipode,* 46(5): 1285–1303.

Caprotti, F., Springer, C. and Harmer, N. (2015) '"Eco" for whom? Envisioning eco-urbanism in the Sino-Singapore Tianjin Eco-city, China', *International Journal of Urban and Regional Research,* 39(3): 495–517.

Carlyle, T. (1824) TC TO ALEXANDER CARLYLE; 14 December 1824, Available at: http://carlyleletters.dukeupress.edu//content/vol3/#lt-18241214-TC-AC-01, (accessed 30 May 2016).

Cartier, C. (2002) 'Transnational urbanism in the reform-era Chinese city: landscapes from Shenzhen', *Urban Studies,* 39(9): 1513–1532.

Carù, A. and Cova, B. (2008) 'Small versus big stories in framing consumption experiences', *Qualitative Market Research: An International Journal,* 11(2): 166 –176.

Castells, M. (1996) *The Rise of the Network Society,* Oxford: Blackwell.

Castells, M. (1977) *The Urban Question: A Marxist Approach,* London: Edward Arnold.

Castells, M. (1976a) 'Is there an urban sociology?', in Pickvance, C. (ed.), *Urban Sociology: Critical Essays,* London: Methuen.

Castells, M. (1976b) 'The wild city', *Kapitalistate,* 4–5: 2–30.

Castells, M. and Portes, A. (1989) 'World underneath: the origins, dynamics, and effects of the informal economy', in Portes, A., Castells, M. and Benton, L. A. (eds) *The Informal Economy: Studies in Advanced and Less Developed Countries,* London: The Johns Hopkins University Press, 66–74

Catungal, J. P., Leslie, D. and Hii, Y. (2009) 'Geographies of displacement in the creative city: The case of Liberty Village, Toronto', *Urban Studies,* 46(5–6): 1095–1114.

Chadwick, E. (1842) 'Report to Her Majesty's Principal Secretary of State for the Home Department', in *the Poor Law Commissioners on an Inquiry into the Sanitary Condition of the Labouring Population of Great Britain with Appendices.* London: W. Clowes and Sons, for HMSO.

Chai, Q. (2002) 'Drawing on land consolidation process, implementing the principle of using land intensively: The formulation and implementation of land replacement policy in Beijing's green belt area' [Lizu tudi zhengli, shishi jiyue yongdi: Beijingshi lvhua geli diqu jianshe tudi zhihuan zhengcede zhiding yu shishi], *Beijing Real Estate [Beijing Fangdichan],* 6: 5–9.

Chakrabarty, D. (2000) *Provincialising Europe: Postcolonial Thought and Historical Difference,* Princeton, NJ: Princeton University Press.

Chakrabarty, D. (1991) 'Open space/public space: garbage, modernity and India', in *South Asia,* 16: 15–31.

Chan, C. K. (2014) 'Constrained labour agency and the changing regulatory regime in China', *Development and Change,* 45: 685–709.

Chan, K. W., Buckingham, W. (2008) 'Is China abolishing the hukou system?', *The China Quarterly,* 195(1): 582–605.

Chang, D-O. (2009) 'Informalising labour in Asia's global factory', *Journal of Contemporary Asia,* 39(2): 161–179.

Chang, L. (2014) 'China world's biggest luxury consumer', *China Daily,* Accessed March 14, 2015, from: http://www.chinadaily.com.cn/business/2014-02/21/content_17298225.htm.

Change of Hutong Culture (n.d.) http://www.ebeijing.gov.cn/Elementals/eBeijing_Neighbourhood/t962476.htm (accessed 17 March 2017)

Chaudhuri, S. (2000) 'Rural-urban migration, the informal sector, urban unemployment, and development policies: a theoretical analysis', *Review of Development Economics*, 4: 353–364.

Chen, B. (2001) *The Cultural History of Shanghai*, Shanghai: China Literature and Art Press.

Chen, B. (2000) *The Grand Sight of Shanghai Anecdotal*, Shanghai: Bookstore Publishing House Press.

Chen, C. (2010) 'Dancing on the streets of Beijing: improved uses within the urban system', in Hou (ed.), *Insurgent Public Space: Guerrilla Urbanism and the Remaking of Contemporary Cities,* New York: Routledge, 12–24.

Chen, F. and Thwaites, K. (2013) *Chinese Urban Design*, Aldershot: Ashgate.

Chen, G. (1996) *Rethinking Beijing: A Memoir [Jinghua daisilu]*, Beijing: Beijing Academy of Urban Planning and Design.

Chen, J. (2000) 'Modern Guangzhou activities in public places-public parks', *Journal of Sun Yatsen University Forum*, 56(20): 116–126.

Chen, J. and Chen, N. (2016) 'Beyond the everyday? Rethinking place meanings in tourism', *Tourism Geographies*, 1–18.

Chen, S. (2009) 'Lu Xun's Neighbours', *Ren Min Zheng Xie Bao*, 23 April 2009: 7.

Chen, Y. (2008) 'The crisis of legitimacy and space of rationality in urban development' [Chengshi kaifade zhengdangxing weiji yu helixing kongjian], *Sociological Studies [Shehuixue Yanjiu],* 23(3): 29–55.

Chen, Y. (1989) 'The control of urban environmental noise', *Chongqing Environmental Science*, 11.1: 59–63. [陈延训：《城市环境噪声及控制》，《重庆环境科学》，1989 (1), 第59–63页。]

Chen, Y. H. (2008) 'On the function of Hangzhou's teahouses as public civic space', *Journal of Hangzhou Normal University,* 5: 116–120. (In Chinese)

Chen, Y. Q. (2004) 'On the change of tourism and entertainment space during the period of late Qing and Republican China', *Shi Lin,* 5: 93–100. (In Chinese)

Cheng, J. (2012a) *Craft of Gardens: The Classic Chinese Text on Garden Design*. Zurich: Enfield.

Cheng, L. (2012b) *Private Gardens: Gardens for the Enjoyment of Artificial Landscapes of Men of Letters*, Shanghai: China Architecture & Building Press.

China Daily (1016) 'China remains largest market for luxury goods', Available at: http://europe.chinadaily.com.cn/business/2016-04/08/content_24372918.htm (Accessed 19 November 2016)

China Internet Information Centre (2001) *Successful Bidding for Hosting the Olympic Games*, China Internet Information Centre, Last Modified August 2001. Available at: http://www.china.org.cn/english/features/38328.htm. (Accessed 15 April 2017).

Christaller, W. (1933) *Central Places in Southern Germany*, London: Prentice Hall

CNNIC (China Internet Network Information Centre) (2007) 'China Blog Market Research Report'; Available at: www.cnnic.cn/uploadfiles/pdf/2007/12/26/113902.pdf (accessed November 2008)

Clarke, A. J. (2001) 'The aesthetics of social aspriration', in Miller, D. (ed.), *Home Possessions: Material Culture Behind Closed Doors*, Oxford: Berg, 23–45

Clough, P. T. (2008) 'The affective turn: political economy, biomedia and bodies', *Theory, Culture and Society,* 25(1): 1741–1758.

Cockain, A. (2011) *Young Chinese in Urban China*, London: Routledge.

Collins, G. and Erickson, A. (2011) 'Dying for a spot: China's car ownership growth is driving a national parking spot shortage', *China Sign Post*, Available at: http://

www.chinasignpost.com/2011/01/10/dying-for-a-spot-chinas-car-ownership-growth-is-driving-a-national-parking-space-shortage/ (Accessed 8 March 2017).

Colomb, C. (2012) 'Pushing the urban frontier: Temporary uses of space, city marketing, and the creative city discourse In 2000s Berlin', *Journal of Urban Affairs*, 34(2): 131–152.

Connell, R. (2007) *Southern Theory: The Global Dynamics of Knowledge in Social Science,* Cambridge: Polity Press.

Conradson, D. and Latham, A. (2007) 'The affective possibilities of London: antipodean transnationals and the overseas experience', *Mobilities,* 2(2): 231–254.

Corbin, Alain. (1998) *Village Bells: Sound and Meaning in the Nineteenth-century French Countryside.* Trans. Martin Thom. New York: Columbia University Press.

Corcuff, S. (2012) 'The liminality of Taiwan: a case-study in geopolitics', *Taiwan in Comparative Perspective,* 4: 34–64.

CPC (1958) 'Resolution on Certain Questions in the People's Commune [Guanyu renmin gongshe ruogan wenti de jueyi]', in *Important Documents since the Founding of PRC: Volume 11 [Jianguo yilai zhongyao wenxian xuanbian: di shiyi juan].* Beijing: Literature Research Center of the CPC Central Committee.

Crang, M. (2001) 'Rhythms of the city: temporalised space and motion', in May, J. and Thrift, N. J. (eds) *TimeSpace: Geographies of Temporality,* New York/London: Routledge, 187–207.

Crang, M. and Zhang, J. (2012) 'Transient dwelling: trains as places of identification for the floating population of china', *Social and Cultural Geography,* 13(8): 895–914.

Cresswell, T. (2010) 'Towards a politics of mobility', *Environment and Planning D: Society and Space,* 28(1): 17–31.

Cresswell, T. (2006a) *On the Move: Mobility in the Modern Western World,* New York: Routledge.

Cresswell, T. (2006b) '"You cannot shake that shimmie here": producing mobility on the dance floor', *Cultural Geographies,* 13: 55–77.

Csikszentmihalyi, M. and Robinson, R. E. (1990) *The Art of Seeing: An Interpretation of the Aesthetic Encounter,* Los Angeles, CA: Getty Publications.

Cui, L. (2012) 'Peaceful rise: China's modernisation trajectory', *The International Spectator: Italian Journal of International Affairs,* 47(2): 14–17.

Cui, Z. H. (2009) 'Theories and researches of modern Chinese parks: on reading cities and parks', *Shi Lin,* 2: 165–172. (In Chinese)

Currie, R. R. (1997) 'A pleasure-tourism behaviours framework', *Annals of Tourism Research,* 24(4): 884–897.

Da Costa, D. (2015) 'Sentimental capitalism in contemporary India: Art, heritage, and development in Ahmedabad, Gujarat', *Antipode,* 47(1): 74–97.

Dai, X. (2015) 'Square dance: ideologies, aesthetic culture and public space', *Journal of Southwest University of Nationalities,* 11: 178–184. (In Chinese)

Dashilar (2016) 'New community in Yangmeizhu: New neighbors in old neighborhood *(Yangmeizhu xin shequn: lao shequ xin linli)*', Available at: http://www.dashilar.org/index.htm#B!/ing/iFresh/t2063.html (Accessed 25 October 2016).

Dashilar (2013a) 'Organizations (*Shishi zhuti*)', Available at: http://www.dashilar.org/index.htm#A!/A/A7a.html (Accessed 25 October 2016).

Dashilar (2013b) 'Dashilar revitalization plan: Strategy *(Dashilar gengxin jihua: celue)*', Available at: http://www.dashilar.org/#A!/A/A2_A.html (Accessed 24 October 2016).

Dashilar (2011) '2011 Dashilar new landscape *(2011 Dashilar xin jiejing)*', Available at: http://www.dashilar.org/#A!/A/A2e.html (Accessed 24 October 2016).

Davidson, J. and Bondi, L. (2004) 'Spatialising affect; affecting space: an introduction', *Gender, Place and Culture,* 11(3): 373–374 .

Davidson, J. O. C. (1998) *Prostitution, Power and Freedom,* Cambridge: Polity Press.

Davidson, J., Smith, M. and Bondi, L. (eds) (2005) *Emotional Geographies,* Aldershot: Ashgate.

Davis, D. (2005) 'Urban consumer culture', *The China Quarterly*, 183(Sep.): 672–709.

Davis, D. (1993) 'Urban households: supplicants to a socialist state', in Davis, D. and Harrell, S. (eds) *Chinese Families in the Post-Mao Era*, Berkeley: University of California Press, 30–44.

Davis, M. (2006) *Planet of Slums*, London: Verso.

Davis, M. (2004) 'Planet of slums: urban involution and the informal proletariat', *New Left Review*, 26: 5–34.

Davis, M. (1990) *City of Quartz: Excavating the Future in Los Angeles*, London: Verso.

De Boeck, F. (2011) 'Inhabiting ocular ground: Kinshasa's future in the light of Congo's spectral urban politics', *Cultural Anthropology*, 26: 263–286.

De Certeau, M. (1984) *The Practice of Everyday Life*, Berkeley: University of California Press.

De Certeau, M., Girard, L. and Mayol, P. (1998) *The Practice of Everyday life, Volume 2: Living and Cooking*, London: University of Minnesota Press.

Deleuze, G. and Guattari, F. (1987) *A Thousand Plateaus: Capitalism and Schizophrenia*, Minneapolis: University of Minnesota Press.

DeLillo, D. (2001) 'Letter to the Translator', in DeLillo, D.m *White Noise* (trans. Zhu, Y.) Nanjing: Yilin Press.

de Soto, H. (1989) *The Other Path: the Invisible Revolution in the Third World*, New York: Harper & Row.

Deutsche, R. and Ryan, C. G. (1984) 'The fine art of gentrification', *October*, 31 (Winter): 91–111.

Dewsbury, J-D. (2000) 'Performativity and the event: enacting a philosophy of difference', *Environment and Planning D Society and Space*, 18: 473–496.

DeVault, M. (1991) *Feeding the Family: The Social Organization of Caring as Gendered Work*, Chicago: University of Chicago Press.

Dickens, L. (2010) 'Pictures on walls? Producing, pricing and collecting the street art screen print', *City*, 14(1–2): 63–81.

Ding, Y. (2012) 'Negotiating intimacies in an eroticized environment: Xiaojies and South China entertainment business', *International Journal of Business Anthropology*, 3(1): 158–175.

Ding, Y., and Ho, P. S.Y. (2013). 'Sex work in China's Pearl River Delta: accumulating sexual capital as a life-advancement strategy', *Sexualities*, 16(1–2): 43–60.

Dixon, D. P. and Straughan, E. R. (2010) 'Geographies of touch/touched by geography', *Geography Compass*, 4(5): 449–459.

Doherty, B. and Doyle, T. (2006) 'Beyond borders: transnational politics, social movements and modern environmentalisms', *Environmental Politics*, 15(5): 697–712.

Dong, L. and Tian, K. (2009) 'The use of Western brands in asserting Chinese national identity', *Journal of Consumer Research*, 36(3): 504–523.

Doody, B. J., Perkins, H. C., Sullivan, J. J., Meurk, C. D. and Stewart, G. H. (2014) 'Performing weeds: gardening, plant agencies and urban plant conservation', *Geoforum*, 56: 124–136.

Downey, G. (2010) 'Practice without theory: a neuro-anthropological perspective on embodied learning', *Journal of the Royal Anthropological Institute*, 16 (s1): S22–S40.

Doyle, T. and McEachern, D. (2001) *Environment and Politics*, London and New York: Routledge.

Drummond, L. B. W. (2000) 'Street scenes: practices of public and private space in urban Vietnam', *Urban Studies*, 37(12): 2377–2391.

Du, L. (201) 'Beijing's Yanxitai builds golf courses illegally in the name of "green belt"' [Beijing yanxitai jie "lvge" mingyi weigui jian gao'erfu qiuchang], *China Business News*, Last Modified 23 July 2011. Available at: http://finance.qq.com/a/20110723/001197.htm. (Accessed 24 March 2017).

Dufrenne, M. (1973) *The Phenomenology of Aesthetic Experience*, New York: Northwestern University Press.

Durkheim, E. (1964 [1893]) *The Division of Labour in Society*, New York: The Free Press.

Eberle, M. L. and Holliday, I. (2011) 'Precarity and political immobilisation: migrants from Burma in Chiang Mai, Thailand', *Journal of Contemporary Asia,* 41(3): 371–392.

Edensor, T. (2012) 'Illuminated atmospheres: anticipating and reproducing the flow of affective experience in Blackpool', *Environment and Planning D: Society and Space*. 30(6): 1103–1122.

Edensor, T. (2010) 'Introduction: thinking about rhythm and space', in Edensor, T. (ed.), *Geographies of Rhythm: Nature, Place, Mobilities and Bodies*, Burlington, VT/Farnham, Surrey, England: Ashgate, 1–18.

Edensor, T. and Holloway, J. (2008) 'Rhythmanalysing the coach tour: the Ring of Kerry, Ireland', *Transactions of the Institute of British Geographers*, 33(4): 483–501.

Edensor, T. and Jayne, M. (eds) (2011) *Urban Theory Beyond the West: A World of Cities,* London: Routledge.

Elfick, J. (2011) 'Class formation and consumption among middle-class professionals in Shenzhen', *Journal of Current Chinese Affairs,* 40(1): 187–211.

Engels, F (1887 [1844]) *The Condition of the Working-Class in England,* London: Swann Sonnenschein.

Evans, D. (2012) 'Beyond the throwaway society: Ordinary domestic practice and a sociological approach to household food waste', *Sociology,* 46(1): 41–56.

Evans, G. (2003) 'Hard-branding the cultural city – from Prado to Prada', *International Journal of Urban and Regional Research*, 27(2): 417–440.

Fang, K. and Zhang, Y. (2003) 'Plan and market mismatch: Urban redevelopment in Beijing during a period of transition', *Asia Pacific Viewpoint,* 44(2): 149–162.

Fang, P. (2006) 'The opera house and an expansion of Shanghai's public space during the late Qing Dynasty', *Journal of East China Normal University,* 38(6): 43–49. (In Chinese)

Färber, A. (2014) 'Low-budget Berlin: towards an understanding of low-budget urbanity as assemblage', *Cambridge Journal of Regions, Economy and Society*, 7(1): 119–136.

Farías, I. (2016) 'Assemblages' in *Urban Theory; New Critical Perspectives*, ed. M. Jayne M and K. Ward, *Urban Theory; New Critical Perspectives*, London: Routledge, 41–51.

Farías, I. (2014) 'Planes maestros como cosmogramas: la articulación de fuerzas oceánicas y formas urbanas en Chile', *Revista Pléyade*, 14: 119–142.

Farías, I. (2011) 'The politics of urban assemblages', *City*, 15 (3–4): 365–374.

Farías, I. (2009) 'Introduction: decentering the object of urban studies', in Farías, I. and Bender, T. (eds), *Urban Assemblages: How Actor-Network Theory Changes Urban Studies*, London: Routledge.

Farías, I. and Bender, T. (eds) (2009) *Urban Assemblages: How Actor-Network Theory Changed Urban Studies*, London: Routledge.

Featherstone, M. (1991) *Consumer Culture and Postmodernism*, London: Sage.

Feng, J. (1990) 'Man and nature – the comparative history of garden and architecture construction development trend', *Journal of Architecture*, 1: 27–34.

Feng, P. M. (2016) 'Development of city parks and modern sports in the late Qing Dynasty and the early Republic of China periods', *Journal of Physical Education,* 23(5): 1–22. (In Chinese)

Ferrer, G. L. (2006) 'The Chinese social dance party in Tokyo: identity and status in an immigrant leisure subculture', *Journal of Contemporary Ethnography*, 33(6): 651–674.

Ferreri, M. (2015) 'The seduction of temporary urbanism', *Ephemera*, 15(1): 181–191.

Florida, R. (2002a) 'Bohemia and economic geography', *Journal of Economic Geography*, 2(1): 55–71.

Florida, R. (2002b) *The Rise of the Creative Class*, New York, NY: Basic Books.

Florida, R. (2002c) 'The rise of the creative class', *Washington Monthly*, May. Available at: https://www.creativeclass.com/rfcgdb/articles/14%20The%20Rise%20of%20the%20Creative%20Class.pdf (Accessed 20 February 2017).

Fong, V. (2011) *Paradise Redefined: Transnational Chinese Students and the Quest for Flexible Citizenship in the Developed World*, Palo Alto, CA: Stanford University Press.

Foucault, M. (2008) *The Birth of Biopolitics: Lectures at the College de France, 1978–79*, London: Palgrave Macmillan.

Foucault, M. (1975) *Discipline and Punish*, New York: Vintage.

Foucault, M. (1979) *The History of Sexuality, Vol. 1*, London: Allen Lane.

Friedmann, J. (2005) *China's Urban Transition*, Minneapolis, MN: University of Minnesota Press.

Friedmann, J. (1986) 'The world city hypothesis', *Development and Change* 17: 69–88.

Frijda, N. H. and Sundararajan, L. (2007) 'Emotion refinement: a theory inspired by Chinese poetics', *Perspectives on Psychological Science*, 2(3): 227–241.

Frijns, J., Phung, T-P. and Mol, A (2000) 'Ecological modernisation theory and industrialising economies. The case of Viet Nam', *Environmental Politics* 9(1): 257–292.

Frith, W. P. (1891) *John Leech: His Life and Work* (Vol. 2), London: Richard Bentley.

Fullagar, S., Markwell, K. and Wilson, E. (eds) (2012) *Slow Tourism: Experiences and Mobilities*, London: Channel View Publications.

Gallagher, M. and Prior, J. (2014) 'Sonic geographies: Exploring phonographic methods', *Progress in Human Geography*, 38(2): 267–284

Gandy, M. and Nilsen, B. (2014) *The Acoustic City*, Cambridge: Jovis.

Gardiner, M. E. (2000) *Critiques of Everyday Life*, London: Routledge.

Gaubatz, P. (2008) 'New public space in urban China: fewer walls, more malls in Beijing, Shanghai and Xining', *China Perspectives,* 4: 72–83.

Gaubatz, P. (1999) 'China's urban transformation: patterns and processes of morphological change in Beijing, Shanghai and Guangzhou', *Urban Studies,* 36: 1495–1521.

Gaubatz, P. (1995) 'Changing Beijing', *Geographical Review*, 85: 79–96.

Geng, J. G. (2001) 'Harm and prevention from home appliance noise', *Home Appliance*, 7: 51.

Gill, R. and Pratt, A. (2008) 'In the social factory?: Immaterial labour, precariousness and cultural work'. *Theory, Culture and Society,* 25(7–8): 1–30.

Gilmartin, C. K. (1994) *Engerndering China: Women, Culture and The State*, Cambridge, MA: Harvard University Press.

Ginn, F. and Demeritt, D. (2009) 'Nature: a contested concept', in Clifford, N. J., Holloway, S. L., Rice, S. P. and Valentine, G. (eds) *Key Concepts in Geography*, 2nd edition, London and Thousand Oaks, CA: Sage, 300–311.

Goldman, M. (2011) 'Speculative urbanism and the making of the next world city', *International Journal of Urban and Regional Research*, 35(3): 555–581.

Goldsmith, M. (2012) *Discord: the Story of Noise*, Oxford: Oxford University Press.

Gomez, M. V. (1998) 'Reflective images: The case of urban regeneration in Glasgow and Bilbao', *International Journal of Urban and Regional Research*, 22(1): 106–121.

Goodman, B. and Larson, W. (2005) *Gender in Motion: Divisions of Labor and Cultural Change in Late Imperial and Modern China*, Lanham, MD: Rowman & Littlefield

Gottlieb, A. (1982) 'Americans' vacations', *Annals of Tourism Research*, 9: 165–187.

Gramsci, A. (1971) *Selections from the Prison Notebooks* (Translated by Geoffrey Nowell Smith and Quintin Hoare), London: Lawrence and Wishart.

Gregg, M. and Seigworth, G. J. (2010) *The Affect Theory Reader*, New York: Duke University Press.

Gregson, N. (2007) *Living with Things: Ridding, Accommodation, Dwelling.* Wantage: Sean Kingston Publishing.

Grodach, C. (2012) 'Cultural economy planning in creative cities: Discourse and practice', *International Journal of Urban and Regional Research*, 37(5): 1747–1765.

Grodach, C., Foster, N. and Murdoch, J. (2016) 'Gentrification, displacement and the arts: Untangling the relationship between arts industries and place change', *Urban Studies*, 29(1): 2131–2412.

Grosz, E. (1999) 'Bodies-cities', in Price, J. and Shildrick, M. (eds) *Feminist Theory and the Body*, Edinburgh: Edinburgh University Press, 24–36.

Guangzhou Daily. (2009) Retrieved August 9, http://news.163.com/09/0807/08/5G3O9N3R000120GR.html

Gui, C. L., Wu, L.Y. and Cook, I. (2012) 'Progress in research on Chinese urbanism', *Frontiers of Architectural Research*, 1(2): 101–149.

Gui, Y., Ma, W. and Muhlhahn, K. (2009) 'Grassroots transformation in contemporary China', *Journal of Contemporary Asia*, 39(3): 400–423.

Hackworth, J. and Smith, N. (2001) 'The changing state of gentrification', *Tijdschrift voor Economische en Sociale Geografie*, 92(4): 464–477.

Hall, P. (2004) 'European cities in a global world', in Eckhardt, F. and Hassenpflug, D. (eds), *Urbanism and Globalization*, Frankfurt: Peter Lang.

Hall, P. (1966) *The World Cities*, New York: McGraw Hill.

Halsey, M. and Pederick, B. (2010) 'The game of fame: Mural, graffiti, erasure', *City*, 14(1–2): 82–98.

Hannam, K., Sheller, M. and Urry, J. (2006) 'Editorial: mobilities, immobilities and moorings', *Mobilities*, 1(1): 1–22.

Hanser, A. (2005) 'The gendered rice bowl: The sexual politics of service work in urban China', *Gender and Society*, 19: 581–600.

Haraway, D. (1991) *Simians, Cyborgs and Women: The Reinvention of Nature*, New York/London: Routledge.

Hardt, M. (1999) 'Affective labour', *Boundary*, 26(2): 89–100

Harris, A. (2005) 'Opening up the symbolic economy of contemporary Mumbai', in Hall, T. and Miles, M. (eds) *Advances in Art and Urban Futures*, Bristol: Intellect Books, 29–41.

Harris, E. (2015) 'Navigating pop-up geographies: Urban space-times of flexibility, interstitiality and immersion', *Geography Compass*, 9(11): 592–603.

Harrison, P. (2000) 'Making sense: embodiment and the sensibilities of the everyday', *Environment and Planning D: Society and Space*, 18: 497–517.

Hart, K. (1973) 'Informal income opportunities and urban employment in Ghana', *Journal of Modern African Studies*, 11: 61–89.

Harvey, D. (1969) *Explanation in Geography*, London: Arnold.

Harvey, D. (2014) *Seventeen Contradictions and the End of Capitalism*, London: Profile Books.

Harvey, D. (2013) 'Rebel Cities: the Current Outlook', Millercom Lecture, University of Illinois at Urbana-Champaign, 13th May.

Harvey, D. (2012) *Rebel Cities: From the Right to the City to the Urban Revolution*, London: Verso.

Harvey, D. (2008) 'The right to the city', *New Left Review*, 53: 23–40.

Harvey, D. (2006) *Spaces of Global Capitalism*, New York: Verso Books.

Harvey, D. (2005) *A Brief History of Neoliberalism*, Oxford: Oxford University Press.

Harvey, D. (2003) *The New Imperialism*, Oxford: Oxford University Press.

Harvey, D. (2002) 'The art of rent: globalization, monopoly and the commodification of culture', *Socialist Register*, 38: 93–110.

Harvey, D. (2000) *Spaces of Hope*, Berkeley: University of California Press.

Harvey, D. (1989a) 'From managerialism to entrepreneurialism:The transformation in urban governance in late capitalism', *Geografiska Annaler: Series B, Human Geography*, 71(1): 3–17
Harvey, D. (1989b) *The Condition of Postmodernity: An Enquiry into the Origins of Cultural Change*, Oxford: Basil Blackwell.
Harvey, D. (1989c) *The Urban Experience*, Oxford: Blackwell.
Harvey, D. (1985) *The Urbanization of Capital*, Baltimore, MD:The Johns Hopkins University Press.
Harvey, D. (1982) *The Limits to Capital*, Oxford: Blackwell.
Harvey, D. (1973) *Social Justice and the City*, Oxford: Blackwell.
Harvey, D. (1969) *Explanation in Geography*, London: Arnold.
Hauser, S. M. and Xie, Y. (2005) 'Temporal and regional variation in earnings inequality: urban China in transition between 1988 and 1995', *Social Science Research*, 34(1): 44–47.
He, J. L. (2011) 'Public space and health in Beijing in the early Republic of China', *Journal of National Museum of China*, 2: 127–134. (In Chinese)
He, S. and Qian, J. (2017) 'New frontiers in researching Chinese cities', in Hannigan, J. and Richards, R. (eds) *The Sage Handbook of New Urban Studies*, London: Sage, 462–478.
He, S. J. (2013) 'Evolving enclave urbanism in China and its socio-spatial implications: the case of Guangzhou', *Social and Cultural Geography*, 14(3): 243–275.
He, S. J. and Qian, J. X. (forthcoming) 'From an emerging market to a multifaceted urban society: urban China studies', *Urban Studies*.
He, S. and Wu, F. (2009) 'China's emerging neoliberal urbanism: perspectives from urban redevelopment, *Antipode*, 41: 282–304.
Heberer, T. and Gobel, C. (2013) *The Politics of Community Building in Urban China*, London: Routledge.
Heidegger, M. (1996) *The Principle of Reason*, Bloomington, IN: Indiana University Press.
Hemmings, C. (2005) 'Invoking affect cultural theory and the ontological turn', *Cultural Studies*, 19: 548–567.
Herschikovitz, L. (1993) 'Tiananmen Square and the politics of place', *Political Geography*, 12(5): 395–420.
Hershatter, G. (2004) 'State of the field: women in China's long twentieth century', *The Journal of Asian Studies*, 63: 991–1005.
Hessler, P. (2006) 'Hutong Karma –The life of a Beijing alleyway', *New Yorker, New Yorker Magazine Incorporated*: 82–89.
Heynen, N., Perkins, H. A., and Roy, P. (2006) 'The political ecology of uneven urban green space: the impact of political economy on race and ethnicity in producing environmental inequality in Milwaukee', *Urban Affairs Review*, 42(1): 3–25.
Ho, P. and Edmond, R. (2007) 'Perspectives of time and change: rethinking embedded environmental activism in China', *China Information*, XXI (2): 331–344.
Ho, S, and Lin, G. (2003) 'Emerging land markets in rural and urban China: policies and practices', *The China Quarterly*, 175(3): 681–707.
Hoffman, S. (2014) 'Escalating land protests in Yunnan', *China Policy Institute: Analysis*. Available at: https://cpianalysis.org/2014/11/06/land-protestspredictablepreventable/ (Accessed 15 October 2017).
Hoffman, L. (2011) 'Urban modelling and contemporary technologies of city-building in China: The production of regimes of green urbanisms', in Roy, A. and Ong, A. (eds) *Worlding Cities: Asian Experiments and the Art of Being Global*, Oxford: Wiley-Blackwell, 55–76.
Hondagneu-Sotelo, P. (2010) 'Cultivation questions for a sociology of gardens', *Journal of Contemporary Ethnography*, 39(5): 498–516.

Horkheimer, M. and Adorno, T. W. (1947) *Dialectic of Enlightenment*, Stanford, CA: Stanford University Press..

Hsing, Y-T (2010) *The Great Urban Transformation: Politics of Land and Property in China*, Oxford: Oxford University Press.

http://assets1c.milkeninstitute.org/assets/Publication/ResearchReport/PDF/China-HousingMarket.pdf (accessed 8 March 2017)

Hu, J. (2013) *Splendid Chinese Garden: Origins, Aesthetics and Architecture*, New York: Better Link Press.

Huang, C.T. (2013) 'A political park: the Working People's Cultural Palace in Beijing', *Journal of Contemporary History,* 48(3): 556–577.

Huang, G. Z. and Xue, D. S. (2009) 'Review of informal employment research in China', *Tropical Geography*, 29: 389–393. (In Chinese with English summary)

Huang, G. Z., Xue, D. S. and Zhang, H. O. (2016) 'The development of urban informal employment and its effects on urbanization in China', *Geographical Research*, 35: 442–454. (In Chinese with English summary)

Huang, J. (2008) *Modern design application of Chinese traditional gardening theory of "Man living in harmony with nature"*, paper presented at the 9th International Conference on Computer-Aided Industrial Design and Conceptual Design, CAID/CD2008.

Huang, P. C. C. (2009) 'China's neglected informal economy: reality and theory', *Modern China*, 35: 405–38.

Huang, P. C. C. (1993) '"Public sphere"/ "civil society" in China? The third realm between state and society', *Modern China,* 19(2): 216–240.

Huang, Y and Li, S-M, (2014) *Housing Inequality in Chinese Cities*, London: Routledge.

Hubbard, P. (2006) *City*, London: Routledge.

Hubbard, P. and Prior, J. (2013) 'Out of sight, out of mind? Prostitution policy and the health, well-being and safety of home-based sex workers', *Critical Social Policy,* 33(1): 140–159.

Hubbard, P., Matthews, R. and Scoular, J. (2008). 'Regulating sex work in the EU: prostitute women and the new spaces of exclusion', *Gender, Place and Culture,* 15(2): 137–152.

Hung, C-F. (2013) 'Citizen journalism and cyberactivism in China's Anti-PX Plant in Xiamen, 2007–2009', *China: An International Journal*, 11(1): 40–54.

Hutton, T. A. (2004) 'The new economy of the inner city', *Cities*, 21(2): 89–108.

Illouz, E. (2009) 'Emotions, imagination and consumption a new research agenda,' *Journal of Consumer Culture*, 9(3): 377–413.

ILO (2002) *Women and Men in the Informal Economy: A statistical picture,* Geneva: International Labour Organization.

IOC (1996) *The Olympic Charter*, Lausanne, Switzerland: International Olympic Committee.

Iossifova, D. (2011) 'Shanghai borderlands: The rise of a new urbanity?' in Edensor, T. and Jayne, M. (eds) *Urban Theory Beyond the West: A World of Cities*, London: Routledge, 195–208

Isin, E. (ed.) (2000) *Democracy, Citizenship and the Global* City, New York: Routledge.

Isoke, Z. (2013) *Urban Black Women and the Politics of Resistance*, Basingstoke: Macmillan.

Jacobs, J. (1992) *The Death and Life of Great American Cities,* New York. Vintage Books.

Jansen, S. (2009) 'After the red passport: Towards an anthropology of the everyday geopolitics of entrapment in the EU's "immediate outside"', *Journal of the Royal Anthropological Institute,* 15: 815–832.

Jayne, M. (2016) 'Sharon Zukin', in Latham, A. and Koch, R. (eds) *Key Thinkers on Cities*, London: Sage.

Jayne, M. (2013) 'Ordinary urbanism –neither trap nor tableaux', *Environment and Planning A*, 45: 2305–2313.

Jayne, M. (2005) *Cities and Consumption*, London: Routledge.

Jayne, M. and Bell, D. (2010) 'Urban order', in Kitchin, R. and Thrift, N. (eds), *The International Encyclopedia of Human Geography*, London: Elsevier.

Jayne, M. and Ferenčuhová, S. (2014) 'Comfort, identity and fashion in the post-socialist city: assemblages, materiality and context', *Journal of Consumer Culture*, 15 (4): 329–350.

Jayne, M. and Leung, H. H. (2014) 'Embodying Chinese urbanism: towards a research agenda', *Area* 46(3): 256–267.

Jayne, M. and Ward, K. (eds) (2017) *Urban Theory: New Critical Perspectives*, London: Routledge.

Jayne, M., Gibson, C., Waitt, G. and Valentine, G. (2012) 'Drunken mobilities: backpackers, alcohol, "doing place"', *Tourist Studies*, 13(8): 211–231.

Jayne, M., Hubbard, P. and Bell, D. (2011) 'Worlding a city: Twinning and urban theory', *City: Analysis of Urban Trends, Culture, Theory, Policy and Action*, 15(1): 25–41.

Jayne, M., Hubbard, P. and Bell, D. (2013) 'Twin cities: territorial and relational geographies of worldly Manchester', *Urban Studies*, 50(2): 239–254.

Jayne, M., Valentine, G. and Holloway, S. L. (2010) 'Emotional, embodied and affective geographies of alcohol, drinking and drunkenness', *Transactions of the Institute of British Geographers*, 35(4): 540–554.

Jiang, J., Rosenqvist, U., Wang, H., Greiner, T., Lian, G. and Sarkadi, A. (2006) ,'Influence of grandparents on eating behaviors of young children in Chinese three-generation Families', *Appetite* , 48(3): 377–383.

Jim, C. Y. and Chen, W.Y. (2006) 'Impacts of urban environmental element on residential housing prices in Guangzhou (China)', *Landscape and Urban Planning* 78: 422–434.

Jing, J (ed.) (2000) *Feeding China's Little Emperors: Food, Children and Social Change*. Stanford, CA: Stanford University Press.

Johnson, I. (2013) 'Leaving the land: Picking death over eviction', *The New York Times*, 8th September. Available at: http://www.nytimes.com/2013/09/09/world/asia/as-chinese-farmers-fight-for-homes-suicide-is-ultimate-protest.html?pagewanted=all (Accessed 15 October 2017).

Johnson, T. (2010) 'Environmentalism and NIMBYism in China: promoting a rules-based approach to public participation', *Environmental Politics*, 19(3): 430–448.

Johnson, L. (2006a) 'Browsing the modern kitchen – a feast of gender, place and culture (part 1)', *Gender, Place and Culture*, 13(2): 123–132.

Johnson, L., Johnson-Pynn, J. and Pynn, T. (2007) 'Youth civic engagement in China: results from a program promoting environmental activism', *Journal of Adolescent Research*, 22(4): 355–386.

Johnston, L. (2009) 'The Body', in Kitchin, R. and Thrift, N. (eds) *The International Encyclopaedia of Human Geography*, London: Elsevier, 326–331.

Joy, A. and Sherry, J. F. (2003) 'Speaking of art as embodied imagination: a multisensory approach to understanding aesthetic experience', *Journal of Consumer Research*, 30(2): 259–282.

Joy, A. and Venkatesh, A. (1994) 'Postmodernism, feminism and the body: the visible and the invisible in consumer research' *International Journal of Research in Marketing*, 11(4): 333–357.

Joy, A., Wang, J. F., Chan, T. S., Sherry, J. F. and Cui, G. (2014) 'M(Art)Worlds: consumer perceptions of how luxury brand stores become art institutions', *Journal of Retailing*, 90(3): 347–364.

Kafka, F. (2013) *Letters to Felice*, New York: Knopf Doubleday Publishing Group.

Kagawa, T and Jinfeng, C. (2007) *Residential Space Change and Restructure in Shanghai Downtown Area; With the Case of Luwan District, Shanghai* (unpublished report).

Kaiman, J. (2012) 'Razing History: The Tragic Story of Beijing Neighborhood's Destruction', *The Atlantic*. Available at: https://www.theatlantic.com/international/archive/2012/02/razing-history-the-tragic-story-of-a-beijing-neighborhoods-destruction/252760/ (Accessed 8 March 2017).

Kantor, P. and Savitch, H.V. (2005) 'How to study comparative urban development politics: a research note', *International Journal of Urban and Regional Research*, 29(2): 135–151.

Kapferer, J. N. (2010) 'Luxury after the crisis: pro logo or no logo', *The European Business Review*, September – October: 42–46.

Katz, C. and Kirby, A. (1991) 'In the nature of things: the environment and everyday life', *Transactions of the Institute of British Geographers*, 16(3): 277–342.

Kempadoo, K. (2001) 'Freelancers, temporary wives, and beach-boys: researching sex work in the Caribbean', *Feminist Review*, 67: 39–62.

Kenworthy-Teather, E. (2001) 'The case of the disorderly graves: contemporary deathscapes in Guangshou', *Social and Cultural Geography*, 2(2): 185–202.

Kenworthy-Teather, E. (1999) *Embodied Geographies: Spaces, Bodies and Rites of Passage*, London: Routledge.

Keizer, G. (2010) *The Unwanted Sound of Everything We Want: A Book About Noise*, New York: Public Affairs.

Keswick, M., Jencks, C. and Hardie, A. (2003) *The Chinese Garden: History, Art and Architecture*, London: Frances Lincoln.

Knapp, R. G. (2000) *China's old dwellings*, Honolulu: University of Hawai'i Press.

Knapp, R. G. and Lo, K-Y. (2005) *House, Home, Family: Living and Being Chinese*. Honolulu: University of Hawai'i Press.

Knox, P. L. (1987) 'The social production of the built environment: architects, architecture and the postmodern city', *Progress in Human Geography* 21: 154–377.

Kong, L. (2011) *Chinese Male Homosexualities: Memba, Tongzhi and Golden Boy*, London: Routledge.

Kong, L. (2010) 'China and geography in the 21st century: a cultural (geographic) revolution', *Eurasian Geography and Economics*, 51(5): 6000–6618.

Kong, L. (2007) 'Cultural icons and urban development in Asia: Economic imperative, national identity, and global city status', *Political Geography*, 26(4): 383–404.

Kong, T. S. K. (2006) 'What it feels like for a whore: The body politics of women performing erotic labour in Hong Kong', *Gender, Work and Organization*, 13(5): 409–434.

Krivý, M. (2011) 'Speculative redevelopment and conservation: The signifying role of architecture', *City*, 15(1): 42–62.

Kucera, D. and Roncolato, L. (2008) 'Informal employment: two contested policy issues', *International Labour Review*, 147: 321–348.

Kwan, S. (2013) *Kinesthetic City: Dance and Movement in Chinese Urban Spaces*, Oxford: Oxford University Press.

Laing, M. and Cook, I. R. (2014) 'Governing sex work in the city', *Geography Compass*, 8(8): 505–515.

LANDESA (2012) *Summary of 2011 17-Province Survey Findings*, Available at: https://www.landesa.org/china-survey-6/ (Accessed 15 October 2017).

Lastovicka, J. L. and Sirianni, N. J. (2011) 'Truly, madly, deeply: consumers in the throes of material possession love', *Journal of Consumer Research*, 38(2): 323–342.

Latour, B. (2005) *Reassembling the Social*, Oxford: Oxford University Press.

Latour, B. (1993) *We Have Never Been Modern*, Hemel Hempstead: Harvester Wheatsheaf.

Lea, J. (2012) 'Negotiating therapeutic touch: encountering massage through the "mixed bodies" of Michael Serres', in Paterson, M. and Dodge, M. (eds) *Touching Space, Placing Touch*, Aldershot: Ashgate, 29–46.

Lee, L. O. F. (1999) *Shanghai Modern: The Flowering of a New Urban Culture in China, 1930–1945*, Cambridge, MA: Harvard University Press.
Lee, N. K. (2009) 'How is a political space made? The birth of Tiananmen Square and the May Fourth movement', *Political Geography,* 28(1): 32–43.
Lefebvre, H. (1968) *La Droit a La Ville*, Paris: Anthropos.
Lefebvre, H. (2004) *Rhythmanalysis: Space, Time and Everyday Life*, London: Continuum.
Lefebvre, H. (2003) *The Urban Revolution*, Minneapolis, MN: University of Minnesota Press.
Lefebvre, H. (1996) *Writings on Cities* (translated by E. Kofman and E. Lebas), Oxford: Blackwell.
Lefebvre, H. (1991a) *Critique of Everyday Life I: Introduction*, New York/London: Verso.
Lefebvre, H. (1991b) *The Production of Space*, Oxford: Basil Blakewell.
Lefebvre, H. (1984) *Critique of Everyday Life*, London: Verso.
Lefebvre, H. (1971) *Everyday Life in the Modern World*, London: Allan Lane.
Lefebvre, H. (1968) *La Droit a La Ville*, Paris: Anthropos.
Lett, J.W. (1983) 'Ludic and liminoid aspects of charter yacht tourism in the British Virgin Islands', *Annals of Tourism Research*, 10: 35–56.
Ley, D. (2011) *Millionaire Migrants: Transpacific Life Lines*, Chichester, UK: John Wiley and Sons.
Ley, D. (2003) 'Artists, aestheticisation and the field of gentrification', *Urban Studies*, 40(12): 2527–2544.
Li, D.Y. (2009) 'Urban public space and social life: a case study of parks in the modern time', *Urban History Research,* (Z2): 127–153. (In Chinese)
Li, K. (2013) 'Assembling consensus, forming collegiality, promoting the process of urbanisation in a more stable and ameliorate way [Ningju gongshi, xingcheng heli, tuidong chengzhenhua gengwen genghao fazhan]', *Selected Party Literature since the 18th National Congress of the Communist Party of China [Shibada yilai zhongyao wenxian xuanbian]*, edited by CCCPC Party Literature Research Office (2016), 608–622. Beijing: CCCPC Party Literature Publishing House.
Li, M. (2009) *30 Talks on the Chinese Classical Gardens*, Beijing: China Architecture and Building Press.
Li, Y. (2012) 'Wildscapes in China: A case study of the Houtan Wetland Park – Expo 2010 Shanghai', in Jorgensen, A. and Keenan, R. (eds) *Urban Wildscapes,* London: Routledge, 111–119.
Lim, J. (2007) 'Queer critique and the politics of affect', in Browne, K., Brown, G. and Lim, J. (eds) *Geographies of Sexuality,* London: Ashgate, 53–67.
Lin, G. (2009) *Developing China: Land, Politics and Social Conditions*, London and New York: Routledge.
Lin, G. (2007) 'Chinese urbanism in question: state, society, and the reproduction of urban spaces', *Urban Geography,* 28(1): 7–29.
Lin, G. and Zhang, A-Y. (2015) 'Emerging spaces of neoliberal urbanism in China: Land commodification, municipal finance and local economic growth in prefecture-level cities', *Urban Studies,* 52(15): 2774–2798.
Lin, G., Li, X., Yang F-W. and Hu, Z-Y. (2015) 'Strategizing urbanism in the era of neoliberalization: State power reshuffling, land development and municipal finance in urbanizing China', *Urban Studies*, 52(11): 1962–1982.
Lin, M. H and Bao, J. G. (2016) 'Research on "carnivalesque dancing": a case study of Guangzhou', *Tourism Tribune,* 31(6): 60–72. (In Chinese)
Liu-Farrer, G. (2016) 'Migration as class-based consumption: the emigration of the rich in contemporary China', *The China Quarterly*, 226 (June): 499–518.
Liu, A. (1994) 'Hazard and prevention from family noise', *Sports World*, 10: 40.
Liu, X (2011) 'A world all its own: a study and exploration of Chinese traditional garden'. *Neighbourhood Spaces, Environment and Behaviour*, 36(5): 678–700.

Lloyd, R. (2002) 'Neo–Bohemia: art and neighborhood redevelopment in Chicago', *Journal of Urban Affairs*, 24(5): 517–532.

Logan, J. F. (2001) *New Chinese City: Globalization and Market Reform*, London: Blackwell.

Longhurst, R. (2001) *Bodies: Exploring Fluid Boundaries*, London: Routledge.

Loram, A., Tratalos, J., Warren, P. H. and Gaston, K. J. (2007) 'Urban domestic gardens: the extent and structure of the resource in five major cities', *Landscape Ecology,* 22: 601–615.

Lorimer, H. (2008) 'Cultural geography: non-representational conditions and concerns', *Progress in Human Geography*, 36(3): 1–9.

Lozada, E. P. (2000) 'Globalized childhood? Kentucky Fried Chicken in Beijing', in. Jing, J., *Feeding China's Little Emperors* (pp. 114–134). Stanford, CA: Stanford University Press.

Lu, A. (2016) 'Owning a car in China is getting harder and harder', *Shanghai Daily.* Available at: http://www.shanghaidaily.com/business/autotalk-special/Owning-a-car-in-China-is-getting-harder-and-harder/shdaily.shtml (Accessed 8 March 2017).

Lu, D. (2011) *Remaking Chinese Urban Form: Modernity, Scarcity and Space, 1949–2005*, London: Routledge.

Lu Xun. (1959) *Lu Xun's Diary* (Volume 1). Beijing: People's Literature Publishing House.

Lu Xun. (1973) *ErYi Ji • Xiao Za Gan*. Beijing: People's Literature Publishing House.

Lu, X. (1973) *ErYi Ji • Xiao Za Gan*', Beijing: People's Literature Publishing House [鲁迅：《而已集•小杂感》，北京：人民文学出版社，1973年版]

Lu, X. (1959) '*Lu Xun's Diary* (Volume 1)', Beijing: People's Literature Publishing House. [鲁迅：《鲁迅日记》（上卷），北京：人民文学出版社，1959年版，第83页。]

Lu, Y. (2005) *Environmental Civil Society and Governance in China*, London: Chatham House.

Luo, S. and Huang, F. (2013) 'China's Olympic dream and the legacies of the Beijing Olympics', *The International Journal of the History of Sport*, 30(4): 443–452.

Ma, C. (2009) *Shanghai Foreign Settlement*, Tianjin: Tianjin Education Press.

Ma, L. J. C. (2005) 'Urban administrative restructuring, changing scale relations and local economic development in China', *Political Geography,* 24(4): 477–497.

Ma, C and Liu, T. (2008) 'Ferocious Noise', *Southern Weekly*, 15 May 2008: C17. [马昌博、柳天伟：《噪声凶猛》，《南方周末》2008–05–15，C17版。]

Ma, C. and Feng, S. (2001) *The Records of Shanghai Settlement*, Shanghai Academy of Social Sciences Press.

Ma, D. (1987) 'The status quo of China's urban noise', *Acta Acustica*, 12(2)： 81–91. [马大猷：《中国城市噪声的现状》，《声学学报》，1987(2)]

Ma, L. (2009) 'Chinese Urbanism', in Kitchin, R. and Thrift, N. (eds) *International Encyclopedia of Human Geography*, Amsterdam: Elsevier.

Ma, L. (2002) 'Urban transformation in China, 1949–2000: a review and research agenda', *Environment and planning A*, 34(9): 1545–1570.

Ma, L. (1976) 'Anti-urbanism in China', *Proceedings of the Association of American Geographers*, 8: 114–118.

Ma, L. J. C. (2006) 'The state of the field of urban China: a critical multidisciplinary overview of the literature', *China Information*, 20(3): 363–389.

Ma, L. J. C. and Wu, F. (2005) *Restructuring the Chinese City: Changing Society, Economy and Space*, New York/London: Routledge.

Madsen, R. (1998) *China's Catholics: Tragedy and Hope in an Emerging Civil Society,* Berkeley: University of California Press.

Mahmood, S. (2011) *Politics of Piety: The Islamic Revival and the Feminist Subject*, Princeton, NJ: Princeton University Press.

Mai, Q and Franchesh-Huidobro, R. (2014) *Climate Change Governance in Chinese Cities*, London: Routledge.

Maldonado, C. (1995) 'The informal sector: legalization or lassez-faire?', *International Labour Review*, 134: 705–728.

Maloney, W. F. (2004) 'Informality revisited', *World Development*, 32: 1159–1178.

Markusen, A. (2006) 'Urban development and the politics of a creative class: Evidence from a study of artists', *Environment and Planning A*, 38(10): 1921–1940.

Martin, D. (2001) 'Food restrictions in pregnancy among Hong Kong mothers', in Wu, D.Y. and Tan, C.-b., *Changing Chinese Foodways in Asia* (pp. 97–122). Hong Kong: The Chinese University Press.

Massey, D. (2004) 'Geographies of responsibility', *Geografiska Annaler* 86 B (1): 5–18.

Massey, D. (1995) 'The Conceptualization of place', in Massey, D. and Jess, P. (eds) *A Place in the World?: Places, Cultures and Globalization*, Milton Keynes: Open University Press, 45–85.

Massey, D. (1993) 'Power-geometry and a progressive sense of place', in Bird, J. (ed.), *Mapping the Futures: Local Cultures, Global Change*, London: Routledge, 60–70.

Massumi, B. (2002) *Parables for the Virtual Movement, Affect, Sensation*, Durham, NC and London: Duke University Press.

Mathews, V. (2014) 'Incoherence and tension in culture-led redevelopment', *International Journal of Urban and Regional Research*, 38(3): 1019–1036.

Mathews, V. (2008) 'Artcetera: Narrativising gentrification in Yorkville, Toronto', *Urban Studies*, 45(13): 2849–2876.

Mazumdar, S. and Mazumdar, S. (2012) 'Immigrant home gardens: places of religions, culture, ecology, and family', *Landscape and Urban Planning*, 105: 258–262.

Mbembe, A. and Nuttall, S. (2004) 'Writing the world from an African metropolis', *Public Culture*, 16: 347–372.

McAuliffe, C. (2012) 'Graffiti or street art? Negotiating the moral geographies of the creative city', *Journal of Urban Affairs*, 34(2): 189–206.

McCann, E. and Ward, K. (2013) 'A multi-disciplinary approach to policy transfer research: geographies, assemblages, mobilities and mutations', *Policy Studies*, 34(1): 2–18.

McCann, E. and Ward, K. (2012) 'Assembling urbanism: following policies and studying through the sites and situations of policy making', *Environment and Planning A*, 44(1): 42–51.

McCann, E. and Ward, K. (eds) (2011) *Mobile Urbanism: Cities and Policymaking in the Global Age*, Minneapolis: University of Minnesota Press.

McCann, E. and Ward, K. (2010) 'Relationality/territoriality: toward a conceptualization of cities in the world', *Geoforum*, 41(3): 175–184.

McCormack, D. P. (2008a) 'Geographies of moving bodies: thinking, dancing, spaces', *Geography Compass*, 2(6): 1822–1836.

McCormack, D. P. (2008b) 'Engineering affective atmospheres on the moving geographies of the 1897 Andrée expedition', *Cultural Geographies,* 15(4): 413–430.

McGee, T. (2011) *China's Urban Space: Development Under Market Socialism*, London: Routledge.

McLean, H. (2014) 'Digging into the creative city: A feminist critique', *Antipode* 46(3): 669–690.

Meah, A. and Jackson, P. (2013) 'Crowded kitchen: The "democratisation of domesticity"?' *Gender, Place and Culture: A Journal of Feminist Geography*,, 20(5): 578–596.

Merrifield, A. (2002) *Metromarxism: A Marxist Tale of the City*, New York: Routledge.

Merrifield, A. (2014a) *The New Urban Question*, London: Pluto Press.

Merriman, P. (2014b) 'Rethinking mobile methods', *Mobilities*, 9(2): 167–187.

Miller, D. (2010) *Stuff,* Cambridge: Polity Press.

Miller, D. (2008) *The Comfort of Things,* Cambridge: Polity Press.

Miller, D. (2002) 'Accommodating', in Painter, C. (ed.), *Contemporary Art and the Home*, Oxford: Berg, 115–130.

Miller, D. (2001a) 'Behind closed doors', in Miller, D. (ed.) *Home Possessions: Material Culture Behind Closed Doors*, Oxford: Berg, 1–19.

Miller, D. (2001b) *Home Possessions: Material Culture Behind Closed Doors,* Oxford: Berg.

Miller, D. (1998) *Material Cultures: Why Some Things Matter,* Chicago: University of Chicago Press.

Miles, M. (2005) 'Interruptions: Testing the rhetoric of culturally led urban development', *Urban Studies*, 42 (5–6): 889–911.

Miles, S. (2007) 'Consumption as freedom: intergenerational relationships in a changing China', in Powell, J. and Cook, I. (eds) *Aging in Asia,* Beijing: Casa Verde, 32–43.

Miles, S. (1998) *Consumerism As A Way of Life,* London: Sage.

Min, I., Fang, C. and Li, Q. (2004) 'Investigation of patterns in food-away-from-home expenditure for China', *China Economic Review,* 15(4): 457–476.

Mitchell, D. (2016) 'Tent cities: interstitial spaces of survival', in Brighenti, A. M. (ed.), *Urban Interstices: The Aesthetics and the Politics of the In-between*, London: Routlede, 67–78.

Mitchell, D. (2003) *The Right to the City: Social Justice and the Fight for Public Space*, New York: Guilford Press.

Mitchell, K. (2004) *Crossing the Neoliberal Line: Pacific Rim Migration and the Metropolis*, New York: Temple.

MLR (2002) *MLR-2002-No.11: Regulations on the Assignment of State-owned Land Use Right by Bidding, Auction and Quotation* [Guotu ziyuan buling 2002 di 11 hao: zhaobiao paimai guapai churang guoyou tudi shiyongquan guiding], edited by Ministry of Land and Resources. Beijing.

MLR (2004) *MLR-2004-No.71: Announcement on Continuing to Carry out Monitoring and Law Enforcement Works on the Leasing of Land Use Rights by Tender, Auction and Quotation* [Guotu zifa 2004 di 71 hao: Guanyu jixu kaizhan jingyingxing tudi shiyongquan zhaobiao paimai guapai churang qingkuang zhifa jiancha gongzuo de tongzhi], Beijing: Ministry of Land and Resources.

Mol, A. (2006) 'Environment and modernity in transitional China; frontiers of ecological modernization', *Development and Change*, 37(1): 29–56.

Molz, J. G. (2009) 'Representing pace in tourism mobilities: staycations, slow travel and the amazing race', *Journal of Tourism and Cultural Change*, 7(4), 270–286.

Mommaas, H. (2004) 'Cultural clusters and the post-industrial city: Towards the remapping of urban cultural policy', *Urban Studies*, 41(3): 507–532.

Mu, M. (2015) 'Teach how to reduce the noise of washing machines and then your irritable feelings', Available at: http://www.hea.cn/2015/0821/229847.shtml, (Accessed 1 May 2016).

Murakami-Wood, D. and Ball, K. (2013) 'Brandscapes of control? Surveillance, marketing and the co-construction of subjectivity and space in neo-liberal capitalism', *Marketing Theory*, 13(1): 47–67.

Nash, C. (2000) 'Performativity in practice: some recent work in cultural geography', *Progress in Human Geography*, 24(4): 653–664.

Naylor, L. (2012) 'Hired gardens and the question of transgression: lawns, food gardens and the business of "alternative" food practice', *Cultural Geographies*, 19(4): 483–504.

Nightingale, A. (2003) 'A feminist in the forest: Situated knowledges and mixing methods in natural resource management', *An International E-Journal for Critical Geographies*, 2(1): 77–90.

NPC (1998) *Land Management Act of People's Republic of China*, edited by The National People's Congress of the People's Republic of China. Beijing.

Oakley, K. (2009) 'The politics of cultural work', *Cultural Trends*, 18(3): 273–274.

O'Brien, K. J. and Deng, Y. (2017) 'Preventing protest one person at a time: Psychological coercion and relational repression in China', *China Review*, 17(2): 179–201.

O'Callaghan, C. and Lawton, P. (2015) 'Temporary solutions? Vacant space policy and strategies for re-use in Dublin', *Irish Geography*, 48(1): 69–87.

O'Connor, J. and Xin, G. (2006) 'A new modernity?', *International Journal of Cultural Studies*, 9(3): 271–283.

Olmedo, C. and Murray, M. J. (2002) 'The formalization of informal/precarious labour in contemporary Argentina', *International Sociology*, 17: 421–443.

Ong, L. H. (2014) 'State-led urbanisation in China: Skyscrapers, land revenue and "concentrated villages"', *The China Quarterly*, 217: 162–179.

Orum, A. M., Bata, S., Li, S. M., Tang, J. W., Sang, Y. and Thrung, N. T. (2009) 'Public man and public space in Shanghai today', *City and Community*, 8(4): 369–389.

Osburg, J. (2013) *Anxious Wealth: Money and Morality Among China's New Rich*, Palo Alto, CA: Stanford University Press.

Ossewaarde, M. (2007) 'Cosmopolitanism and the society of strangers', *Current Sociology*, 55(3): 367–388.

Pang, J. C. (2015) 'Tactics and practice of building new public entertainment space during the Minguo period', *Journal of Zhejiang Normal University*, 40(4): 63–71. (In Chinese)

Pang, L. (2008) 'China who makes and fakes: a semiotics of the counterfeit', *Theory, Culture and Society*, 25(6): 117–140.

Park, R. E. (1915) 'The City: suggestions for the investigation of human behaviour in the city environment', *American Journal of Sociology*, 20(5): 577–612.

Park, R. E., Burgess, E. W. and McKenzie, R. D. (1925) *The City*, Chicago: University of Chicago Press.

Parnell, S. and Robinson, J. (2012) '(Re)theorizing cities from the global South: looking beyond neo-liberalism', *Urban Geography*, 33(4): 593–617.

Paterson, M. (2005) 'Affecting touch: towards a "felt" phenomenology of therapeutic touch', in Davidson, J., Bondi, L. and Smith, M. (eds) *Emotional Geographies*, Aldershot: Ashgate, 161–173.

Paterson, M., Dodge, M. and Mackian, S. (2012) 'Introduction: Placing touch within social theory and empirical study', in Paterson, M. and Dodge, M. (eds) *Touching Space, Placing Touch*, Aldershot: Ashgate, 1–28.

Pauleit, S. (2006) 'Perspectives on urban Greenspace in Europe', *Journal of Built Environment*, 29(2): 33–47.

Pauleit, S., Slinn, P., Handley, J. and Lindley, S. (2003) 'Promoting the natural greenstructure of towns and cities: English nature's accessible natural greenspace standards model', *Journal of Built Environment*, 29(2): 157–170.

Payer, P. (2007) 'The age of noise: early reactions in Vienna, 1870-1914', *Journal of Urban History*, 33 (5): 773–9.

Pearce, P.L. (2009) 'Now that is Funny: humour in tourism settings', *Annals of Tourism Research*, 36(4): 627–644.

Pearce, P. L. and Pabel, A. (2015) *Tourism and Humour*, Bristol/Buffalo/Toronto: Channel View Publications.

Peck, J. (2007) 'Banal urbanism: cities and the creativity fix', *Magazine on Urbanism*, 7: 36–47.

Peck, J. (2005) 'Struggling with the creative class', *International Journal of Urban and Regional Research*, 29(4): 740–770.

Pei, L. (2003) 'Cultural connotations of the "funeral dance" of Tujia Nationality of Western Hubei', *Journal of Hubei Institute for Nationalities*, 6: 1–21.

Penêda, V. (2012) 'Pajama karma: Fashion trend says pajama's are chic on the street, but are they?', *Metro Beijing section, Global Times*. Available at: http://www.globaltimes.cn/content/694897.shtml (Accessed 8 March 2017).

Perry, E. (2008) 'Permanent rebellion? Continuities and discontinuities in Chinese protest', in O'Brien, K. (ed.), *Popular Protest in China,* Cambridge, MA: Harvard University Press, 22–33.

Perry, G. E. and Maloney, F. M. (2007) 'Overview: informality: exit and exclusion', in Perry, G. E., Maloney, F. N., Arias, O. S., Fajnzylber, P., Mason, A. D. and Saavedra-Chanduvi, J. (eds) *Informality: Exit and Exclusion,* Washington, D.C.: World Bank, 36–47.

Picker, M. J. (2000) 'The soundproof study: Victorian professionals, work space and urban noise', *Victorian Studies*, 42(3): 427–453.

Pile, S. (2010) 'Emotions and affect in recent human geography', *Transactions of the Institute of British Geographers*, 35: 5–20.

Pingali, P. (2007) 'Westernization of Asian diets and the transformation of food systems: Implications for research and policy', *Food Policy* , 32(3): 281–298.

Plaza, B. (2008) 'On some challenges and conditions for the Guggenheim Museum Bilbao to be an effective economic re-activator', *International Journal of Urban and Regional Research,* 32(2): 506–517.

Podmore, J. (1998) '(Re)Reading the "loft living" habitus in Montreal's inner city', *International Journal of Urban and Regional Research*, 22(2): 283–302.

Pomfret, J. (2013) *China Village Seethes Over Land Grabs as Beijing Mulls New Laws*, Reuters, 7 March. Available at: http://www.reuters.com/article/us-china-landreform/china-village-seethes-over-land-grabs-as-beijing-mulls-new-laws-idUSBRE9260CH20130307 (Accessed 15 October 2017).

Portes, A. (1997) 'Neoliberalism and the sociology of development: emerging trends and unanticipated facts', *Population and Development Review*, 23: 229–259.

Pow, C-P. (2012) 'China exceptionalism: unbounding narratives on urban China', in Edensor, T. and Jayne, M. (eds) *Urban Theory Beyond the West: A World of Cities,* London: Routledge, 47–65.

Pow, C. P. and Neo, H. (2015) 'Modelling green urbanism in China', *Area,* 47(2): 132–140.

Pozini, D. and Rossi, U. (2010) 'Becoming a creative city: The entrepreneurial mayor, network politics and the promise of an urban renaissance', *Urban Studies*, 47(5): 1037–1057.

Pratt, G. (2005) 'Abandoned women and spaces of the exception', *Antipode,* 37(5): 1052–1078.

Prince, R. (2013) 'Consultants and the global assemblage of culture and creativity', *Transactions of the Institute of British Geographers*, 39(1): 90–101.

Pritchard, A. and Morgan, N. (2006) 'Hotel Babylon? Exploring hotels as liminal sites of transition and transgression', *Tourism Management,* 27: 762–772.

Pun, N. and Lu, H. (2010) 'Unfinished proletarianization: self, anger, and class action among the second generation of peasant-workers in present-day China', *Modern China*, 36: 493–519.

Qian, J. X. (2014a) 'Public space in non-Western contexts: practices of publicness and the socio-spatial entanglement', *Geography Compass,* 8(11): 834–847.

Qian, J. X. (2014b) 'From performance to politics? Constructing public and counterpublic in the singing of red songs', *European Journal of Cultural Studies,* 17(5): 602–628.

Qian, J. X. (2014c) 'Performing the public man: cultures and identities in China's grassroots leisure class', *City and Community,* 13(1): 26–48.

Qiu, J. (2005) 'The internet in China: technologies of freedom in a statist society', in Castells, M. (ed.), *The Network Society: A Cross-Cultural Perspective,* Cheltenham: Edward Elgar, 99–124.

Radley, A. (1995) 'The elusory body and social construtionist theory', *Body and Society,* 1 (3): 23–34.

Rafferty, K. (2011) 'Class-based emotions and the allure of fashion consumption', *Journal of Consumer Culture*, 11(2): 239–260.

Rakowski, C. A. (1994) 'Convergence and divergence in the informal sector debate: a focus on Latin America, 1984–92', *World Development*, 22: 501–516.

Rauchfleisch, A. and Schafer, M. (2015) 'Multiple public spheres of Weibo: a typology of forms and potentials of online public spheres in China', *Information, Communication and Society*, 18(2): 139–155.

Raven, J. H., Chen, Q., Tolhurst, R. J. and Garner, P. (2007) 'Traditional beliefs and practices in the postpartum period in Fujian Province, China: A qualitative study', *BMC Pregnancy and Childbirth*, 7(8): 1–11.

Rawes, P. (2007) *Irigaray for Architects*, London: Routledge.

Reinach, S. S. (2005) 'China and Italy: Fast fashion versus prêt a porter – towards a new culture of fashion', *Fashion Theory*, 9(1): 43–56.

Ren, X. (2013) *Urban China*, Cambridge: Polity Press.

Report on China's Environmental Noise Pollution and Control, The (2016) http://dqhj.mep.gov.cn/dqmyyzshjgl/zshjgl/201608/t20160830_363286.shtml

Revill, G. (2016) 'How is space made in sound? Spatial mediation, critical phenomenology and the political agency of sound', *Progress in Human Geography*, 40(2): 240–256.

Revill, G. (2004) 'Cultural geographies in practice: performing French folk music: dance, authenticity and nonrepresentational theory', *Cultural Geographies*, 11: 199–209.

Richard, L. (1991) *La Vie quotidienne sous la République de Weimar*, Paris : Hachette.[里昂耐尔•理查尔：《魏玛共和国时期的德国（1919–1933）》，李末译，济南：山东画报出版社，2005年版(Chinese version)

Richards, E. J. (1961) 'Noise annoyance and its assessment', in Haslett, A. W. and St. John, J. (eds), *Science Survey* 2. London: Vista Books, Longacre Press Ltd, 1–10.

Richaud, L. (forthcoming) 'Between "face" and "faceless" relationships in China's public places: ludic encounters and activity-oriented friendships among middle- and old-aged urbanites in Beijing public parks', *Urban Studies*.

Riding, A. (1993) '2000 Olympics Go to Sydney In Surprise Setback for China', *New York Times*, 24 September. http://www.nytimes.com/1993/09/24/sports/olympics-2000-olympics-go-to-sydney-in-surprise-setback-for-china.html (Accessed 15 October 2017).

Roberts, J. (2006) *Philosophizing the Everyday: Revolutionary Praxis and the Fate of Cultural Theory*, Ann Arbor, MI/London: Pluto Press.

Roberts, T. (2012) 'From "new materialism" to "machinic assemblage": agency and affect in IKEA', *Environment and Planning A*, 44(10): 2512–2529.

Robinson, J. (2016) 'Thinking cities through elsewhere: tactics for a more global urban studies', *Progress in Human Geography* (online early view).

Robinson, J. (2013) '"Arriving at" urban policies/the urban: traces of elsewhere in making city futures', in O. Söderström (ed.), *Critical Mobilities*, London: Routledge.

Robinson, J. (2011) 'Cities in a world of cities: the comparative gesture', *International Journal of Urban and Regional Research*, 35(3): 1–23.

Robinson, J. (2006) *Ordinary Cities: Between Modernity and Development*, New York/London: Routledge.

Robinson, J. (2002) 'Global and world cities: a view from off the map', *International Journal of Urban and Regional Research*, 26(3): 531–554.

Rock, M. Y. (2012) 'Splintering Beijing: Socio-spatial fragmentation, commodification and gentrification in the *hutong* neighborhoods of "old" Beijing' (Doctoral dissertation) University Park, PA: Pennsylvania State University.

Rofel, L. (2007) *Desiring China: Experiments in Neoliberalism, Sexuality, and Public Culture*. Durham, NC: Duke University Press.

Rofel, L. (1992) 'Rethinking modernity: space and factory discipline in China', *Cultural Anthropology*, 7: 93–114.
Rose, G. (1993) *Feminism and Geography: The Limits of Geographical Knowledge*, Minneapolis: University of Minnesota Press.
Rothenberg, J. and Lang, S. (2015) 'Repurposing the High Line: Aesthetic experience and contradiction in West Chelsea', *City, Culture and Society*, 12(2): 12–26.
Rowe, W.T. (1990) 'The public sphere in modern China', *Modern China,* 16(3): 309–329.
Rowe, W.T. (1984) *Hankow: Commerce and Society in a Chinese city, 1796–1889*, Stanford, CA: Stanford University Press.
Roy, A. (2016) 'What is urban about critical urban theory?' *Urban Geography,* 37(6): 810–823.
Roy, A. (2015) 'Governing the postcolonial suburbs', in Hamel, P. and Keil, R. (eds) *Suburban Governance: A Global View*, Toronto: University of Toronto Press.
Roy, A. (2013) 'Book Review Forum: learning the city', *Urban Geography*, 34(1): 131–134.
Roy, A. (2011a) 'Conclusion: Postcolonial urbanism: Speed, hysteria, mass dreams', in Roy, A. and Ong, A. (eds), *Worlding Cities*, Oxford: Wiley-Blackwell.
Roy, A. (2011b) 'Slumdog cities: rethinking subaltern urbanism', *International Journal of Urban and Regional Research*, 35(2): 223–238.
Roy, A. (2009) 'The 21st century metropolis: new geographies of theory', *Regional Studies,* 43: 819–830.
Roy, A. (2003) 'Paradigms of propertied citizenship: transnational techniques of analysis', *Urban Affairs Review*, 38: 463–490.
Roy, A. and Ong, A. (2011) (eds), *Worlding Cities: Asian Experiments and the Art of Being Global*, Oxford: Wiley Blackwell.
Russell, C. A. and Levy, S. (2012) 'The temporal and focal dynamics of volitional experiences', *Journal of Consumer Research*, 39(2): 341–359.
Russo, Alexander. (2009) 'An American Right to an 'Unannoyed Journey'? Transit Radio as a Contested Site of Public Space and Private Attention, 1949–1952'. *Historical Journal of Film, Radio, and Television*, 29(1): 1–25.
Sai, H. (2015) 'In the Old Days, Enjoy a Bottle of Father Tea', *Hainan Daily.* Haikou, http://hnrb.hinews.cn/html/2015-10/14/content_20_2.htm (Accessed 20 November 2016).
Salter, M. B. (2003) *Rights of Passage,* Boulder, CO: Lynne Rienner.
Sandstrom, U. G., Angelstam, P. and Mikusinski, G. (2006) 'Ecological diversity of birds in relation to the structure of urban green space', *Landscape and Urban Planning*, 77: 39–53.
Sanya City Bureau of Statistics (2015) *Sanya Statistic Yearbook*. (Translated from Chinese by China Statistics Press).
Sanyal, B. (1988) 'The urban informal sector revisited: some notes on the relevance of the concept in the 1980s', *Third World Planning Review*, 10: 65–83.
Sargeson, S. (2016) 'Grounds for self-government? Changes in land ownership and democratic participation in China', *The Journal of Peasant Studies,* 2(6): 29–36.
Sargeson, S. (2013) 'Violence as development: Land expropriation and China's urbanization', *The Journal of Peasant Studies,* 40(6): 1063–1085.
Sassen, S. (2006) *Territory, Authority, Rights: From Medieval to Global Assemblages*, Cambridge, MA: Cambridge University Press.
Sassen, S. (ed.) (2002) *Global Networks, Linked Cities*, London: Routledge.
Sassen. S. (1997) *Informalisation in Advanced Market Economies*, Geneva: International Labour Organization.
Sassen, S. (1994) *Cities in a World Economy*, Thousand Oaks, CA: Pine Forge Press.
Savitch, H. and Kantor, P. (2004) *Cities in the International Marketplace: The Political Economy of Urban Development in North America and Western Europe,* Princeton, NJ: Princeton University Press.

Sayer. A. (2011) *Why Things Matter to People,* New York: Cambridge University Press.

Schaller, S. and Guinand, S. (2017) 'Pop-up landscapes: A new trigger to push up land value?', *Urban Geography*, OnlineFirst

Schmid, C (2012) 'Henri Lefebvre, the right to the city, and the new metropolitan mainstream', in Brenner, N., Marcuse, P. and Mayer, M. (eds), *Cities for People, Not for Profit: Critical Urban Theory and the Right to the City*, London: Routledge, 42–63.

Schneider, A., Seto, K.C. and Webster, D. (2005) 'Urban growth in Chengdu, western China: application of remote sensing to assess planning and policy outcomes', *Environment and Planning B*, 32(3), 323–334.

Schopenhauer, A. (2000) 'On din and noise', *Parerga and Paralipomena* (Vol. 2), Oxford: Clarendon Press.

Schuetz, A. (1944) 'The stranger: an essay in social psychology', *American Journal of Sociology,* 49(6): 499–507.

Scott, A. and Storper, M. (2015) 'The nature of cities: The scope and limits of urban theory', *International Journal of Urban and Regional Research*, 39(1): 1–15.

Scott, A. J. (2006) 'A perspective of economic geography', *Journal of Economic Geography*, 4(5): 479–499.

Scott, A. J. (ed) (2001) *Global City-Regions: Trends, Theory, Policy,* Oxford: Oxford University Press.

Scott, A. J. (1998) *Regions in the World Economy: The Coming Shape of Global Production, Competition and Political Order,* Oxford : Oxford University Press.

Scott, A. J., Agnew, J., Soja, E. W. and Storper, M. (2001) 'Global city-regions', in Scott, A. J. (ed.), *Global City-Regions: Trends, Theory, Policy*, Oxford: Oxford University Press.

Seto, K. C. (2004) 'Urban growth in South China: winners and losers of China's policy reforms', *Geographische Mitteilungen*, 148: 50–57.

Shapiro, J. (2001) *Mao's War Against Nature: Politics and the Environment in Revolutionary China,* Cambridge: Cambridge University Press.

Shaqiao, L. (2014) *Understanding the Chinese City*, London: Sage.

Sheller, M. and Urry, J. (2006) 'The new mobilities paradigm', *Environment and Planning A*, 38(2): 207–226.

Shen, Y. (2013) *IFamily: individual, family and nation of China's urbanisation (in Chinese).* [沈奕斐.个体家庭iFamily：中国城市现代化进程中的个体、家庭与国家. 上海: 三联书店] Shanghai: Shanghai Joint Publication Press.

Sherry, J. F. (2005) 'Brand meaning', in T. Calkins and A. Tybout (eds), *Kellogg on Branding*, Hoboken, NJ: John Wiley, 40–69.

Sherry, J. F. (1998) 'The soul of the company store: Nike Town Chicago and the emplaced brandscape', in Sherry, J. F. (ed.), *Servicescapes: The Concept of Place in Contemporary Markets*, Chicago, IL: NTC Business Books, 109–150.

Sherry, J. F. (1995), 'Anthropology of marketing: retrospect and prospect', in Sherry, J. F. (ed.), *Contemporary Marketing and Consumer Behavior: An Anthropological Sourcebook*, Thousand Oaks, CA: Sage, 435–445.

Shi, G. F. (2016) 'Beijing parks opening and new public space development in the early period of Minguo', *Academic Exploration,* 3: 132–136. (In Chinese)

Shi, M. Z. (1998) 'From imperial gardens to public parks: the transformation of urban space in early twentieth-century Beijing', *Modern China,* 24(3): 219–254.

Shields, R. (1992) 'A truant proximity: presence and absence in the space of modernity', *Environment and Planning D: Society and Space,* 10(2): 181–198.

Shillington, L. J. (2013) 'Right to food, right to the city: Household urban agriculture and socio-natural metabolism in Managua, Nicaragua', *Geoforum*, 44: 103–111.

Shillington, L. (2008) 'Being(s) in relation at home: socio-natures of patio "Gardens" in Managua, Nicaragua', *Social and Cultural Geography*, 9(7): 775–776.

Shin, H. B. (2018) 'Geography: Rethinking the "urban" and urbanization', in Iossifova, D., Doll, C. and Gasparatos, A. (eds), *Defining the Urban: Interdisciplinary and Professional Perspectives*, London: Routledge, 27–39.

Shin, H. B. (2014) 'Contesting speculative urbanisation and strategising discontents', *City: Analysis of Urban Trends, Culture, Theory, Policy, Action,* 18(4–5): 509–516.

Shin, H. B. (2013) 'China's Speculative Urbanism and the Built Environment', The China Policy Institute at the University of Nottingham, Last modified 24 April 2013. Available at: https://cpianalysis.org/2013/04/24/chinas-speculative-urbanism-and-the-built-environment/. (Accessed 27 July 2017).

Shin, H. B. (2012) 'Unequal cities of spectacle and mega-events in China', *City,* 16(6): 728–744.

Shin, H. B. (2010) 'Urban conservation and revalorisation of dilapidated historic quarters: The case of Nanluoguxiang in Beijing', *Cities*, 27: S43–S54.

Shin, H. B. (2009a) 'Life in the shadow of mega-events: Beijing Summer Olympiad and its impact on housing', *Journal of Asian Public Policy*, 2(2): 122–141.

Shin, H. B. (2009b) 'Residential redevelopment and the entrepreneurial local state: The implications of Beijing's shifting emphasis on urban redevelopment policies', *Urban Studies* 46(13): 2815–2839.

Shkuda, A. (2015) 'The artist as developer and advocate: real estate and public policy in SoHo, New York', *Journal of Urban History*, 41(6): 999–1016.

Shoup, Y. and Jiang, R. (2016) 'Charging for parking to finance public services', *Journal of Planning Education and Research,* 1(4): 12–26.

Shove, E. (2003) *Comfort, Cleanliness and Convenience: The Social Organization of Normality,* Oxford: Berg.

Shove, E., Watson, M., Hand, M. and Ingram, J. (2007) *The Design of Everyday Life,* Oxford: Berg.

Siegmann, K. A. (2016) 'Understanding the globalizing precariat: from informal sector to precarious work', *Progress in Development Studies*, 16: 111–123.

Sima, Y. (2011) 'Grassroots environmental activism and the Internet: constructing a green public sphere in China', *Asian Studies Review*, 35(4): 477–497.

Sima, Y. and Pugsley, P. (2010) 'The Rise of a "me culture" in postsocialist China: youth, individualism and identity creation in the blogosphere', *The International Communication Gazette*, 72(3): 287–306.

Simmel, G. (1950a) 'The metropolis and mental life', in Wolff, K. (ed.), *The Sociology of Georg Simmel*, London: Collier-Macmillan.

Simmel, G. (1950b) *The Sociology of Georg Simmel*, New York: The Free Press.

Simmel, G. (1908) *Soziologie: Untersuchungen über die Formen der Vergesellschaftung*, Leipzig: Duncker and Humblot.

Simmel, G. (1903) 'The metropolis and mental life', in Kasinitz, P. (ed.), *Metropolis: Center and Symbol for Our Times*, New York: New York University Press.

Simone, A. (2011) *City Life from Jakarta to Dakar: Movements at the Crossroads,* London: Routledge.

Simone, A. (2004) *For the City Yet to Come: Changing African Life in Four Cities*, London: Duke University Press.

Simone, A. (2001) 'Straddling the divides: remaking associational life in the informal African city', *International Journal of Urban and Regional Research*, 25(2): 102–117.

Sit, V. F. S. (2010) *Chinese City and Urbanism: Evolution and Development,* New York: World Scientific.

Smilor, R. W. (1978) 'Confronting the Industrial Environment: the Noise Problem in America, 1893–1932', unpublished thesis, Austin: University of Texas.

Smilor, R. W. (1971) 'Cacophony at 34th and 6th: the noise Problem in America, 1900–1930', *American Studies*, 18(1): 23–28.

Smith, N. (1996) *The New Urban Frontier: Gentrification and the Revanchist City*, London and New York: Routledge.

Smith, R. J. and Hetherington, K. (2013) 'Urban rhythms: mobilities, space and interaction in the contemporary city', in Smith, R. J. and Hetherington, K. (eds), *Urban Rhythms: Mobilities, Space and Interaction in the Contemporary City*, Chichester, West Sussex: John Wiley and Sons, 4–16.

Söderström, O. (2017) 'Mobilities', in Jayne, M. and Ward, K. (eds), *Urban Theory: New Critical Perspectives*, London: Routledge, 193–204.

Sofaer, J. (ed.) (2007) *Material Identities*, Oxford: Blackwell.

Soja, E. (2010) *Seeking Spatial Justice*, Minneapolis: Minnesota University Press.

Soja, E. (1996) Thirdspace: Journeys to Los Angeles and Other Real-and-Imagined Places, Cambridge, MA: Blackwell.

Soja, E. (1989) *Postmodern Geographies: The Reassertion of Space in Critical Social Theory*, London: Verso.

Sonnenfeld, D. A. (2000) 'Contradictions of ecological modernization: pulp and paper manufacturing in South East Asia', *Environmental Politics*, 9(1): 235–256.

State Council (2004) *SCGO-2004-No. 1: Notice of the General Office of the State Council on Suspending the Construction of New Golf Courses* [Guobanfa 2004–1 hao: Guowuyuan bangong ting guanyu zanting xinjian gao'erfu qiuchang de tongzhi], Beijing: General Office of the State Council.

State Council (2000) *SCL-2000-No. 1: Reply and Approval of the State Council on the Master Plan of Land Use in Beijing* [Guohan 2000–1 hao: Guowuyuan guanyu beijingshi tudi liyong zongti guihua de pifu], Beijing: The State Council.

Steinhardt, C. and Wu, F. (2015) 'In the name of the public: environmental protest and the changing landscape of popular contention in China', *The China Journal*, 75: 1–12.

Steinmüller, H. and Tan, T. (2015) 'Like a virgin? Hymen restoration operations in contemporary China', *Anthropology Today*, 31(2): 15–18.

Stephens, A. C. (2015) 'The affective atmospheres of nationalism', *Cultural Geographies*, 6(4): 27–34.

Stern, M. J. and Seifert, S. C. (2010) 'Cultural clusters: The implications of cultural assets agglomeration for neighborhood revitalization', *Journal of Planning Education and Research*, 29(3): 262–279.

Stevens, Q. (2017) 'Play', in Jayne, M. and Ward, K. (eds) *Urban Theory: New Critical Perspectives*, London: Routledge, 218–230.

Stevens, Q. (2007) *The Ludic City: Exploring the Potential of Public Spaces*, London: Routledge.

Storper, M. (1997) *The Regional World*. New York: Guilford Press.

Storper, M. and Scott, A. J. (2009) 'Rethinking human capital, creativity and urban growth', *Journal of Economic Geography*, 9: 147–167.

Strand, D. (1989) *Rickshaw Beijing: City People and the Politics in the 1920s*, Berkeley, CA: University of California Press.

Sun, Q. (2014a) 'Discontented with the neighbor's noise, a man in Tianjin wounded the person and was sentenced to jail', *Today Evening Paper*, Available at: http://news.163.com/14/0711/14/A0SMV0JM00014AED.html, (Accessed 1 May 2015).

Sun, S. (2014b) 'Unbearable noise drove a man to shoot the people dancing in square and disperse them by his Tibetan mastiff', *Beijing Times*, Available at: http://news.jinghua.cn/353/c/201404/27/n3902549.shtml?fm=114la, (Accessed 1 May 2015).

Supski, S. (2006) '"It was another skin": The kitchen as home for Australian post-war immigrant women', *Gender, Place and Culture*, 13(2): 29–37.

Swenson, J. J. and Franklin, J. (2000) 'The effects of future urban development on habitat fragmentation in the Santa Monica Mountains', *Landscape Ecology,* 15: 713–730.

Tan, Q. H. (2014) 'Postfeminist possibilities: unpacking the paradoxical performances of heterosexualized femininity in club spaces', *Social and Cultural Geography,* 15(1): 23–48.

Tang, W. and Parish, W. L. (2000) *Chinese Urban Life Under Reform: The Changing Social Contract,* Cambridge: Cambridge University Press.

Tang, X. (2002) 'The anxiety of everyday life in post-revolutionary China', in Highmore, B. (ed.), *The Everyday Life Reader,* London/New York: Routledge, 67–78.

Taylor, P. (2013) *Extraordinary Cities: Millennia of Moral Syndromes, World-Systems and City/State Relations,* London: Edward Elgar.

Taylor, P. (2004) *World City Network: A Global Urban Analysis,* London: Routledge.

Taylor, P. J., Ni, P., Derruder, B., Hoyler, M., Huang, J. and Wilcox, J. (eds) (2012) *Global Urban Analysis: A Survey of Cities in Globalization,* London: Routledge.

Theiss, J. M. (2004) *Disgraceful Matters: The Politics of Chastity in Eighteenth-century China,* Berkeley: University of California Press.

Thibaud, J.-P. (2003) 'The sonic composition of the city', in Bull, M. and Back, L. (eds) *The Auditory Culture Reader,* Amsterdam: Berg Publishers, 329–342.

Thien, D. (2005) 'After or beyond feeling A consideration of affect and emotion in geography', *Area,* 37(4): 450–456.

Thompson, C. J. (1997) 'Interpreting consumers: a hermeneutical framework for deriving marketing insights from the texts of consumers' consumption stories', *Journal of Marketing Research,* 34(4): 438–455.

Thompson, C. J. and Hirschman, E. C. (1995) 'Understanding the socialized body: a post-structuralist analysis of consumers' self-conceptions, body images, and self-care practices', *Journal of Consumer Research,* 22(2): 139–153.

Thompson, Emily. (2002) *The Soundscape of Modernity: Architectural Acoustics and the Culture of Listening in America, 1900–1933.* Cambridge, MA: MIT Press.

Thompson, L. (1990) 'The influence of experience on negotiation performance', *Journal of Experimental Social Psychology,* 26(6): 528–544.

Thrift, N. (2014) 'The "sentient" city and what it may portend', *Big Data and Society,* 1(1): 20–41.

Thrift, N. (2007) *Non-representational Theory: Space, Politics, Affect,* London: Routledge.

Thrift, N. (2004) 'Movement-space: the changing domain of thinking resulting from the development of new kinds of spatial awareness', *Economy and Society,* 33(4): 582–604.

Thrift, N. (2004a) 'Driving in the city', *Theory, Culture and Society,* 21: 41–59.

Thrift, N. (2004b) 'Intensities of feeling: towards a spatial politics of affect', *Geografiska Annaler: Series B, Human Geography,* 86: 57–78.

Thrift, N. (2000) 'Not a straight line but a curve, or, cities are not mirrors of modernity', in Bell, D. and Haddour, A. (eds), *City Visions.* Harlow: Prentice Hall.

Thrift, N. (1997) 'The still point: resistance, expressive embodiment and dance', in Pile, S. and Keith, M. (eds), *Geographies of Resistance,* London: Routledge, 124–151.

Thrift, N. (1993) 'An urban impasse?', *Theory, Culture and Society,* 10(2): 229–238.

Throop, C. J. (2010) 'Latitudes of loss: on the vicissitudes of empathy', *American Ethnologist,* 37(4): 771–782.

Throop, C. J. (2003) 'Articulating experience', *Anthropological Theory,* 3(2): 219–241.

Tian, T. and Ma, J. (2016) 'How powerful "Chinese Drama" is? A square dance App got 20 million dollars investment', *The Paper,* Available at: http://www.thepaper.cn/newsDetail_forward_1567620, (Accessed 24 November 2016).

Tolia-Kelly, D. (2006) 'Affect–an ethnographic encounter? Exploring the universalist imperative of emotional/affectual geographies', *Area,* 38: 213–17.

Tomba, L. (2005) 'Residential space and collective interest formation in Beijing's housing disputes', *The China Quarterly*, 184(1): 934–951.

Tönnies, F. (1955) *Community and Association (Gemeinschaft und Gesellschaft)*, London: Routledge and Kegan Paul.

Tönnies, F. (1887) *Community and Society*, New York: Harper & Row.

Tsang, A. and Mozur, P. (2017) *Airbnb's Rivals in China Hold Hands in a Nervous Market*. NYTimes, Business Day. Available at: https://nyti.ms/2nB8LV4 (Accessed 22 March 2017).

Tumbat, G. and Belk, R.W. (2010) 'Marketplace tensions in extraordinary experiences', *Journal of Consumer Research*, 38(1): 42–61.

Turner, V. (1979) 'Frame, flow and reflection: Ritual and drama as public liminality', *Japanese Journal of Religious Studies*, 6(4): 465–499.

Turner, V. (1969) *The Ritual Process*, New York: Aldine.

TVBS (2013) 'Discontented with the neighbor's noise, an elderly man in Taiwan killed a famous guitar craftsman', *Taiwan's China Times*, Available at: http://www.taiwan.cn/taiwan/roll/201406/t20140630_6426210.htm, (Accessed 1 May 2015).

Tyler, P. E. (1993) 'There's No Joy in Beijing as Sydney Gets Olympics', *New York Times*, 24 September 1993. Available at: http://www.nytimes.com/1993/09/24/sports/olympics-there-s-no-joy-in-beijing-as-sydney-gets-olympics.html (Accessed 15 October 2017).

UN-Habitat (2009) *Planning Sustainable Cities: Global Report on Human Settlements 2009*, London: Earthscan.

UNESCO (2012) 'Creative Cities Network: Beijing', Available at: http://en.unesco.org/creative-cities//node/98 (Accessed 16 March 2017).

Urry, J. (2007) *Mobilities*, Cambridge: Polity.

Üstüner, T. and Holt, D. B. (2007) 'Dominated consumer acculturation: the social construction of poor migrant women's consumer identity projects in a Turkish squatter', *Journal of Consumer Research*, 34(1): 41–56.

Valentine, G. (2008) 'Living with difference: reflections on geographies of encounter', *Progress in Human Geography*, 32(3): 323–337.

Veblen, T. (1899) *The Theory of the Leisure Class*, London: Constable.

Venkatraman, M. and Nelson, T. (2008) 'From servicescape to consumptionscape: a photo-elicitation study of Starbucks in the new China,' *Journal of International Business Studies*, 39(6): 1010–1026.

Venn, C. (2010) 'Individuation, relationality, affect: rethinking the human in relation to the living', *Body and Society*, 16(1): 129–161.

Victorian London-Populations-Census-total population of London (n.d)

Vom Nervösen (1910) *Die Zeit*, 25 November 1910: 13. http://www.victorianlondon.org/population/population.htm, accessed on 1 September 2016.

Wagner, U. (1977) 'Out of time and place: Mass tourism and charter trips', *Ethnos*, 42: 38–52.

Wang, B. (2013) 'Rethinking the intrinsic value of Hainan Father Tea', *Hainan Today Magazine* 62013, 42–43.

Wang, D. (2008) *The Teahouse: Small Business, Everyday Culture, and Public Politics in Chengdu, 1900–1950*, Stanford, CA: Stanford University Press.

Wang, D. (2003) *Street Culture in Chengdu: Public Space, Urban Commoners, and Local Politics, 1870–1930*, Stanford, CA: Stanford University Press.

Wang, D. (2001) 'Teahouse and social life of Chinese cities in the early twentieth century: taking Chengdu as an example', *Historical Research*, 5: 41–53. (In Chinese)

Wang, D. (1998) 'Street culture: public space and urban commoners in late-Qing Chengdu', *Modern China*, 24(1): 34–72.

Wang, D. and Chen, L. (2015) 'Discontented with upstairs noise from a grandmother who is looking after her granddaughter, a man killed this old neighbor', *The Beijing News*, Available at: http://www.bjnews.com.cn/news/2015/04/08/359280.html, (Accessed 1 May 2015).

Wang, H. (2004) 'Regulating transnational flows of people: An institutional analysis of passports and visas as a regime of mobility', *Identities: Global Studies in Culture and Power*, 11: 351–376.

Wang, J., and Chen, L. (forthcoming) 'Guerrilla warfare, flagship project: spatial politics in practicing Chinese rock in Shenzhen's post-political making of a musical city', *Geoforum*.

Wang, J., and Li, S-M. (Online first) 'State territorilisation, neoliberal governmentality: The remaking of Dafen Oil Painting Village, Shenzhen, China', *Urban Geography*.

Wang, J., Joy, A. and Sherry, J. F. (2013) 'Creating and sustaining a culture of hope: Feng shui discourses and practices in Hong Kong', *Journal of Consumer Culture*, 13(3): 241–263.

Wang, J., Oakes, T. and Yang, Y. (eds) (2016) *Making Cultural Cities in Asia: Mobility, Assemblage, and the Politics of Aspirational Urbanism*, London: Routledge.

Wang, M. (2015) *On Domestic Electric Objects*, Zhengzhou: He Nan University Press.

Wang, Q. (2011) *Chinese Gardens: Understanding Traditional Chinese Architecture*, China Electric Power Press.

Wang, W. (2008) 'Modern parks in Beijing and the development of public space', *Social Sciences of Beijing*, (2): 52–57. (In Chinese)

Wang, Y. P. (2000) 'Housing reform and its impacts on the urban poor in China', *Housing Studies*, 15: 845–864.

Wang, Y. and Murie, A. (1999) 'Commercial housing development in urban China', *Urban Studies*, 36(9): 1475–1494.

Ward, K. (2010) 'Towards a relational comparative approach to the study of cities', *Progress in Human Geography*, 34(1): 471–487.

Ward, K. (2009) 'Commentary: towards a comparative (Re)turn in urban studies? Some reflections', *Urban Geography*, 29: 1–6.

Ward, K. (2007) 'Geography and public policy: activist, participatory and policy geographies', *Progress in Human Geography*, 31: 695–705.

Ward, K. (2006) 'Policies in motion, urban management and state restructuring: the trans-local expansion of Business Improvement Districts', *International Journal of Urban and Regional Research*, 30: 54–70.

Ward, K. (2005) 'Geography and public policy: a recent history of "policy relevance"', *Progress in Human Geography*, 29(3): 310–319.

Wasserstrom, J. (2009) 'Middle-class mobilization', *Journal of Democracy*, 20(3): 29–32.

Watson, B. (1958) *The Biography of Ssu Ma Ch'ien . Ssu Ma Ch'ien Grand Historian Of China*, New York: Columbia University Press.

Watson, M. (2013) 'Food waste', in Jackson, P., *Food Words: Essays in Culinary Cullture*, London: Blomsbury, 244–245

Watson, M., and Meah, A. (2013) 'Food, waste and safety: negotiating conflicting social anxieties into the practices of domestic provisioning', *The Sociological Review*, 60(S2): 102–120.

Watson, S. (2006) *City Publics: The (Dis)enchantments of Urban Encounters*, London: Routledge.

Weiping, W. and Gaubatz, P. (2013) *The Chinese City*, London: Routledge.

Wegner, D. M. (2002) *The Illusion of Conscious Will*, Cambridge, MA: The MIT press.

Wengraf, E. (1911) 'Das Recht auf Lärm', *Die Zeit*, 30 April 1911, 1–2.

White, G., Howell, J. and Shang, X. (1996) *In Search of Civil Society: Market Reform and Social Change in Contemporary China*, Oxford: Clarendon Press.

Whitehead, J. W. R. and Gu, K. (2006) 'Research on Chinese urban form: retrospect and prospect', *Progress in Human Geography*, 30(3): 337–355.

Whitehead, M. (2009) 'The wood for the trees: ordinary environmental injustice and the everyday right to urban nature', *International Journal of Urban and Regional Research*, 33(3): 662–681.

Whitehead, M. (2005) 'Between the marvellous and the mundane: everyday life in the socialist city and the politics of the environment', *Environment and Planning* D: *Society and Space*, 23(3): 273–294.

Whitman, W. (1993) 'To a locomotive in Winter', in Whitman, W. (ed.) *Leaves of Grass*, New York: Modern Library.

Whitson, R. (2007) 'Beyond the crisis: economic globalization and informal work in urban Argentina', *Journal of Latin American Geography*, 6: 121–136.

Williams, R. (1977) *Marxism and Literature*, Oxford and New York: Oxford University Press.

Williams, C. C. and Round, J. (2007) 'Rethinking the nature of the informal economy: some lessons from Ukraine', *International Journal of Urban and Regional Research*, 31: 425–441.

Williams, C. C. and Round, J. (2010) 'Explaining participation in undeclared work', *European Societies*, 12: 391–418.

Wilson, H. F. (2016) 'Encounter', in Jayne, M. and Ward, K. (eds) *Urban Theory: New Critical Perspectives,* London: Routledge, 109–121.

Wirth, L. (1938) 'Urbanism as a way of life', *American Journal of Sociology,* 44(1): 1–24.

Wolkowitz, C. (2002) 'The social relations of body work', *Work, Employment and Society*, 16(3): 497–510.

Wong, N. (2015) 'Advocacy coalitions and policy change in China: a case study of anti-incinerator protest in Guangzhou', *Voluntas*, May, 23–34.

Wong, L. (2011) 'Chinese migrant workers: rights attainment deficits, rights consciousness and personal strategies', *The China Quarterly*, 208: 870–892.

Woodward, I. (2011) 'Towards an object-relations theory of consumerism: The aesthetics of desire and the unfolding materiality of social life', *Journal of Consumer Culture*, 11(3): 366–384.

Woolley, H. (2010) *Living Green: Evidence to Support the Provision of Green Space around Social Housing*, London: Natural England and the Neighbourhoods Green Partnership.

Wu, C. P. (2009) 'Change of public space and urban daily life: taking modern tea house in Nanjing as an example', *Journal of University of Science and Technology Beijing,* 25(3): 1–6.

Wu, F. (2009) *Globalization and the Chinese City,* London: Routledge.

Wu, F. (2004) 'Transplanting cityscapes: the use of imagined globalization in housing commodification in Beijing', *Area*, 36(3): 227–234.

Wu, F. (1997) 'Urban restructuring in China's emerging market economy: Towards a framework for analysis', *International Journal of Urban and Regional Research*, 21(4): 640–663.

Wu, F. (1995) 'Urban processes in the face of China's transition to a socialist market economy', *Environment and Planning C: Government and Policy*, 13(2): 159–177.

Wu, F., Xu, J. and Gar-On Yeh, A. (2007) *Urban Development in Post-reform China: State, Market and Space*, London and New York: Routledge.

Wu, F. L., Zhang, F. Z. and Webster, C. (2013) 'Informality and the development and demolition of urban villages in the Chinese peri-urban area', *Urban Studies*, 50: 1919–1934.

Wu, H. (2005) *Remaking Beijing: Tiananmen Square and the Creation of a Political Space,* Chicago, IL: The University of Chicago Press.

Wu, W. and Gaubatz, P. R. (2012) *The Chinese City*, Abingdon, Oxon/New York, NY: Routledge.

Xie, D. and Chen, Y. (1966) 'Report on investigations of Chongqing Dwelling's noise and sound insulation', *Acta Acustica,* 3: 170–171

Xing, G. X. (2011) 'Urban workers' leisure culture and the "public sphere": a study of the transformation of the Workers' Cultural Palace in reform-era China', *Critical Sociology,* 37(6): 817–835.

Xinhua News Agency (2001) *The Memorabilia of Beijing's Bidding for the Games of the XXIX Olympiad in 2008* [Beijing shenban 2008 nian aoyunhui dashiji], Xinhua News Agency, Last modified 14 July 2001. Available at:. http:// people.com.cn/GB/shizheng/252/5934/5935/20010714/511469.html. (Accessed 15 April 2017).

Xu, J. and Yeh, A G. O. (2009) 'Decoding urban land governance: State reconstruction in contemporary Chinese cities', *Urban Studies* 46(3): 559–581.

Xu, X., Duan, X., Sun, H., and Sun, Q. (2011) 'Green space changes and planning in the capital region of China', *Environmental Management*, 47: 456–467.

Xu, Y. (2000) *The Chinese City in Space and Time: The Development of Urban Form in Sozhou*, University of Hawaii Press.

Xue, D. S. and Huang, G. Z. (2015) 'Informality and the state's ambivalence in the regulation of street vending in transforming Guangzhou, China', *Geoforum*, 62: 156–165.

Yablon, Nick. (2007) 'Echoes of the city: spacing sound, sound-ing space, 1888–1908', *American Literary History*, 19(3): 629–660.

Yan, G. (2013) *Research on the Root of Hainan Culture*, Beijing: Social Sciences Academic Press.

Yang, G. (2009) *The Power of the Internet in China: Citizen Activism Online,* New York: Columbia University Press.

Yang, G. (2003a) 'Weaving a green web: the internet and environmental activism in China', *China Environment Series*, 6: 89–93.

Yang, G. (2003b) 'The internet and civil society in China: A preliminary assessment', *Journal of Contemporary China*, 12(36): 453–475.

Yang, J. (2014a) 'The politics of affect and emotion: Imagination, pontentiality and anticipation in East Asia', in Yang, J. (ed.), *The Political Economy of Affect and Emotion in East Asia,* Abingdon, Oxon and New York: Routledge, 41–74.

Yang, J. (ed.) (2014b) *The Political Economy of Affect and Emotion in East Asia,* Abingdon, Oxon and New York: Routledge.

Yang, J. (2012) *The Bund*, Shanghai: People Publishing Press.

Yang, J. and Zhou, J. (2007) 'The failure and success of greenbelt program in Beijing', *Urban Forestry and Urban Greening*, 6(4): 287–296.

Yang, L., Zhu, J. and Xiong, R. (2003) 'The Chinese modern shallow concession garden-case study in Tianjin and Shanghai', *Journal of Beijing Forestry University* (Social Science Edition), 2(1): 17–21.

Yang, X. (2009) 'Beijing's Greenbelt Area: policy review and remaining problems', [Beijing shiqu lvhua geli diqu zhengce huiguyu shishi wenti], *Journal of Urban and Regional Planning [Chengshiyu quyu guihua yanjiu]* (1): 171–183.

Yang, Y. (2006) 'On the characteristics and cultural connotations of Zhaotong's folk si-tong drum dance', *Journal of Zhaotong Teacher College*, 3: 1–18.

Yen, S. T., Fang, C.and Su, S.-J. (2004) 'Household food demand in urban China: A censored system approach', *Journal of Comparative Economics* , 32(3): 564–585.

Yeoh, B. S. A. (2015) 'Affective practices in the European city of encounter', *City,* 19(4) ; 545–551.

Yeoh, B. S. A. (1996) *Contesting Space: Power Relations and the Urban Built Environment in Colonial Singapore*, Kuala Lumpur: Oxford University Press.

Yong, G., Ma, W. and Muhlhahn, K. (2009) 'Grassroots transformation in contemporary China', *Journal of Contemporary Asia*, 39(3): 400–423.

Yu, G. (1994) 'Are your ears especially valuable?', in Lin, X. (ed.), *Selected Essays of Yu Guangzhong*, Guangzhou: Guangzhou Press, 243–245.

Yu, L. (2014a) *Chinese City and Regional Planning Systems*, Aldershot: Ashgate.

Yu, L. (2014b) *Consumption in China: How China's New Consumer Ideology is Shaping the Nation,* Cambridge: Polity.

Yu, M. (2004) 'A supplement agreement to No.33 document is under preparation, and the land market in Beijing is waiting' [33 hao wenjian buchong xieyi zhengzai niding, beijing tudi jinru jingmoqi], *Beijing Youth Daily*, Last modified 12 February 2004. Available at: http://news.xinhuanet.com/house/2004-02/12/content_1311671.htm. (Accessed 5 April 2017).

Zerah, M-H. (2007) 'Conflict between green space preservation and housing needs: The case of the Sanjay Gandhi National Park in Mumbai', *Cities*, 24(2): 122–132.

Zhang, A.Y. (2016) 'Arts districts or art-themed parks: arts districts repurposed by/for Chinese governments', in Wang, J., Oakes, T. and Yang, Y. (eds), *Making Cultural Cities in Asia: Mobility, Assemblage, and the Politics of Aspirational Urbanism*, London: Routledge, 69–79.

Zhang, D. (2016) *Chinese Courtyard Housing Under Socialist Market Economy.* China Research Centre.Volume 15, No. 1. Available at: http://www.chinacenter.net/2016/china_currents/15-1/chinese-courtyard-housing-under-socialist-market-economy/ (Accessed 8 March 2017)

Zhang, J (2016) 'Which one is noise, square dance or rock and roll?' *The Paper*, Available at: http://www.thepaper.cn/newsDetail_forward_1581808, (Accessed 17 December 2016).

Zhang, J. (2010) 'Features of Shanghai modern plebeian novels from the perspective of public space', *Journal of Shandong University,* 3: 1–10. (In Chinese)

Zhang, J. (2001) *Fifty Years' Planning and Building of Beijing [Beijing guihua jianshe wushinian]*, Beijing: China Bookstore Publishing House.

Zhang, J. J. (2013) 'Borders on the move: cross-strait tourists' material moments on "the other side" in the midst of rapprochement between China and Taiwan', *Geoforum*, 48: 94–101.

Zhang, L. (2010) *In Search of Paradise: Middle-Class Living in a Chinese Metropolis,* Cornell University Press.

Zhang, L. and Ong, A. (2008) *Privatizing China: Socialism from Afar,* Ithica, NY: Cornell University Press.

Zhang, l., Wu, J., Yu, Z. and Shu, J. (2004) 'A GIS-based gradient analysis of urban landscape pattern of Shanghai metropolitan area, China', *Landscape and Urban Planning,* 69: 1–16.

Zhang, T. J. (2006) 'From conventional private garden to modern urban park: Hankow Zhongshan Park (1928~1938)', *Huazhong Architecture*, 24(10): 177–181. (In Chinese)

Zhang, X. D. (2008) *Postsocialism and Cultural Politics: China in the Last Decade of the Twentieth Century,* Durham, NC: Duke University Press.

Zhang, Y. (2007) 'Creating a green and livable Beijing: On the plan and construction of the second green belt area' [Chuangjian lvse yiju beijing: Beijing di'er dao lvhua geli diqu de guihua yu jianshe], *Beijing City Planning and Construction Review [Beijing guihua jianshe]*, (6): 88–91.

Zhao, Y. (2016) *Invisible Green Belts in Beijing: From Romantic Landscape to Businesses Opportunity*, LSE Blog on HEIF Metropolitan Green Belt, Last modified 17 June 2016. Available at: http://www.lse.ac.uk/geographyAndEnvironment/research/GreenBelt/Green-Belt-Blog/Green-Belt-Blog-Home/Invisible-green-belts-in-Beijing.aspx. (Accessed 17 June 2016).

Zheng, T. (2009) *Red Lights: The Lives of Sex Workers in Postsocialist China*, Cambridge: Cambridge University Press.

Zhou, J. (2008) 'The extending of urban public space and social modernization: a case study of Beijing', *Journal of Beijing Union University,* 6(1): 45–49. (In Chinese)

Zhou, N. and Belk, R. W. (2004) 'Chinese consumer readings of global and local advertising appeals', *Journal of Advertising*, 33(3): 63–76.

Zhou, Q. (2004) 'Property rights and land requisition system: choice for China's urbanization', [Nongdi Chanquanyu Zhengdi Zhidu], *China Economic Quarterly*, 4(1): 193–210.

Zhou, W. (1999) *Chinese Traditional Gardening History*, Beijing: Qinghua University Press.
Zhou, X and Chen, Z. (2009) *The History of Shanghai Public Parks*, Shanghai: Tongj University Press.
Zhou, Y. X. (2010) *Urban Geography: Selections of Zhou Yixing's Work,* Beijing: The Commercial Press. (In Chinese)
Zhou, Y. and Lin, J. (2010) 'Urban residential kitchen evolution and future trends', *Arts and Design*, 11: 20–25.
Zhu, C., Zhu, Y., Lu, R., He, R. and Xia, Z. (2012) 'Perceptions and aspirations for car ownership among Chinese students attending two universities in the Yangtze Delta, China', *Journal of Transport Geography,* 24: 315–325.
Zhu, H. and Wei, H. (2016) 'Think globally and act locally: voices of Chinese urban geographers in the international arena', *Journal of Geographical Sciences*, 26(8): 1001–1018.
Zhu, J. (2012) *The Modern Era Garden History of China: Part One*, Beijing: China Building Industry Press.
Zhu, J. (2000) 'Urban physical development in transition to market: the case of China as a transitional economy', *Urban Affairs Review,* 36(2): 178–196.
Zukin, S. (1998) 'Urban lifestyles: diversity and standardisation in spaces of consumption', *Urban Studies*, 35(5/6): 825–840.
Zukin, S. (1989) *Loft Living*, New Brunswick, NJ: Rutgers University Press.

INDEX

Page numbers in *italics* indicate an illustration, **bold** a table and n an endnote